D1279913

101 INCREDIBLE EXPERIMENTS FOR THE SHED SCIENTIST

Hours of Fun for All The Family!

5 7 9 10 8 6

Published in 2006 by Ebury Press, an imprint of Ebury Publishing

A Random House Group Company

Copyright © Quid Publishing Ltd 2006

Rob Beattie has asserted his right to be identified as the author of this Work
in accordance with the Copyright, Designs and Patents Act 1988

All rights reserved. No part of this publication may be reproduced, stored in a retrieval
system, or transmitted in any form or by any means, electronic, mechanical, photocopying,
recording or otherwise, without the prior permission of the copyright owner

The Random House Group Limited Reg. No. 954009

Addresses for companies within the Random House Group can be found at
www.randomhouse.co.uk

A CIP catalogue record for this book is available from the British Library

The Random House Group Limited supports the Forest Stewardship Council® (FSC®), the leading international
forest certification organisation. All our titles that are printed on Greenpeace approved FSC® certified paper carry
the FSC® logo. Our paper procurement policy can be found at www.randomhouse.co.uk/environment

To buy books by your favourite authors and register for offers visit www.randomhouse.co.uk

Conceived, designed and produced by
Quid Publishing
Level Four
Sheridan House
114 Western Road
Hove BN3 1DD
England
www.quidpublishing.com

Author: Rob Beattie
Illustrations: Steven Bannister
Design: Lindsey Johns
Cover image courtesy of Corbis

ISBN-10: 0091914205
ISBN-13: 9780091914202

NOTE
Every effort has been made to ensure that all information contained in this book is correct and compatible
with national standards at the time of publication. This book is not intended to replace manufacturers' instructions
on the use of their tools and products – always follow their safety guidelines.

The author, publisher and copyright holder assume no responsibility for any injury, loss or damage
caused or sustained as a consequence of the use and application of the contents of this book.

101 INCREDIBLE EXPERIMENTS FOR THE SHED SCIENTIST

*Fascinating Fun With
Everyday Objects*

EBURY
PRESS

ROB BEATTIE

CONTENTS

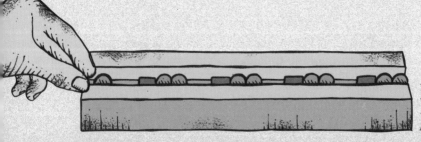

Project 93
Automatic
Pinball,
page 112

*Project 101
The Solar
System in
Your Shed,
page 122*

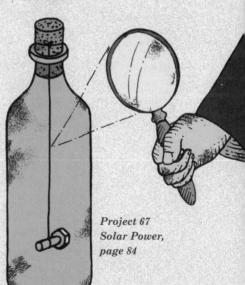

*Project 67
Solar Power,
page 84*

INTRODUCTION

Although the shed has a long and noble history as the last retreat of the contemplative man, it also provides the natural setting for his more practical endeavours. Many have entered the world of the shed seeking rest and relaxation, only to emerge in a fevered state as shed scientists (*Sheddae scientificus*).

When we speak of practical endeavours, these need not necessarily be practical in the most pedantic sense of the word (i.e. being useful or having a purpose), but rather in a broader, more philosophical sense. The shed is where a man (and a woman, for the scientific world should allow no bias on the basis of gender) can debate and demonstrate many of the great truths that underpin our understanding of the physical world.

A shed is at once private (assuming your shed has a door – if it hasn't, build one immediately) and yet easy to get to (if your shed is not conveniently sited in your garden, consider moving it there). Moreover, friends, family and neighbours expect strange sounds and other peculiar emissions from a well-used shed and are thus more accepting of the odd curious odour or muffled bang.

But we jest where we should not. Shed scientists are serious people who share the same lofty ideals as the best-equipped, most highly paid scientists working in a sparkling new laboratory. Never mind that their chosen tools are more likely to be vinegar and baking soda than '2 hydroxy benzoic acid' and 'polytetrafluoroethylene' and that many of their experiments rely on 9-volt batteries and old lollipop sticks rather than centrifuges and spectrophotometers. They are all brothers and sisters in the quest for knowledge.

It is thus a mistake to think of one as the poor relation of the other. Shed scientists may lack the financial advantages of conventional scientists, but that only serves to make them more cunning and even more determined to find out exactly what it is that makes things tick.

Long considered a retreat for gardening types, the shed has evolved along with the shed scientist until the two are a perfect match.

Along the way you'll grow to love the smell of vinegar (which seems to be about one step away from being the most versatile substance on earth) and re-learn a love for ice-lollies.

Don't be overly concerned if you never paid attention in science lessons at school. While rigorous of intent (and certainly very serious about safety – see pages 14 to 16), true shed scientists are a forgiving bunch, more interested in sharing the joys of discovery than setting you extra homework.

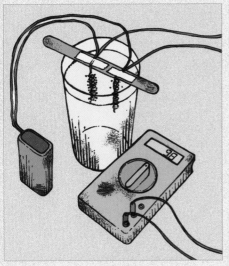

One of the many experiments contained in the pages that follow – a hydrogen fuel cell which produces electricity but no pollution.

Hopefully this book will go a small way towards revealing the secret life of the shed scientist – those lonely, dogged seekers of truth, hunched over a pinhole camera, reeking of home-made insecticide, rubbing balloons on their heads, lit eerily by gherkins with electrodes stuck in them. Noble pastimes all.

In the pages that follow you'll find many experiments that have earned a special place in the pantheon of shed science. Amongst other things you'll learn how to strip uncooked eggs of their shells, to extract DNA from food, turn water into wine, how to grow a garden made of crystals, how to make invisible ink, a Morse-code transmitter, a hovercraft big enough to carry a grown person (or several small ones), a volcano and a hydrogen fuel cell.

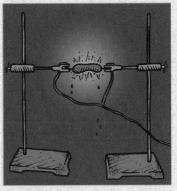

Yes, you too can make ordinary vegetables fizz and pop and glow in the dark.

Physics, pressure, surface tension, balloons, straws – and even candy chromatography. They're all part of shed science!

1 THE SHED LABORATORY

Come on into the shed. You're in good company – many famous laboratories of the past have a shed-like quality about them. On celluloid and in literature, giants of science like Von Frankenstein and Dr. Jekyll built their labs in secret, usually with whatever materials came to hand – something the shed scientist knows about all too well.

Just because there's an element of 'make-do' about shed science, it doesn't mean you should scrimp on the essentials, and it'll make your life easier and your science more productive if you kit out your lab so that it's comfortable and serves your purpose. We understand that unless you're particularly dedicated, shed science usually has to fight for space with other, equally important jobs and pastimes. It's important therefore

Fig. 1. *The simple pleasure that is a tidy workbench. Spend a little time squaring everything away at the end of each experiment.*

that your 'lab' or lab area is discrete (i.e. separate, rather than well-behaved), well organised and kept tidy.

Those of you who read *101 Things to Do in a Shed* will be familiar with the concept of the shed as a state of mind and that, really, anywhere can be a shed. Sadly, this is less true of the Shed Laboratory, and although you'll be able to do many of the experiments described in these pages in a spare bedroom or a home office, with others… well, you're just asking for trouble. Here, the kitchen is your only real alternative and this will require understanding partners (spouses, housemates, pets, and so on) and a commitment to cleaning up as you go.

Assuming you have an actual shed to work in, some of the basic decisions are going to be forced on you. If you've got mains electricity running into your shed, you've got a huge advantage because you'll have all the light and power you need to run various bits of equipment. It means you can scavenge up those 'specialist devices' that you'll need occasionally, like a small fridge, a hotplate and even a transformer.

If you haven't got electricity, all is not lost, but it does mean you'll have to spread your lab load into the house – and possibly negotiate the necessary access with other people – and this may complicate things. You may have to become a diplomat as well as a scientist, or at least

Fig. 2. *Some people see a shed – the resourceful would-be scientist sees a setting for experimental triumphs.*

coincide your raids into the kitchen with those times when other people are out of the house.

Obviously, a lab can be as basic or sophisticated as money, time and circumstances allow. Here we suggest three different versions.

Fig. 3. *Adding mains power to your shed broadens the kind of experiments you can do and makes it more suitable as a late-night lab.*

The 1950s shed lab will:
- Be a well-lit, well-ventilated shed that you can lock when you need to.
- Have a clear work area or bench where you can conduct experiments.
- Include somewhere to store your equipment safely, out of reach.
- Have a working first-aid kit – which you know how to use.
- Have something to put fires out with.

The 1970s version will add:
- Some kind of semi-permanent lighting that allows you to work into the evening – this can be powered by bottled gas, oil or one of those energy-saving hand-cranked jobs.
- A camping stove that uses bottled gas for heating your experiments (and the occasional cup of tea).
- A large cold box for storing occasional perishable ingredients.
- Poor-quality, gaudy decorations.

The 21st-century shed lab will add:
- Mains electricity properly 'plumbed' in by an electrician and fitted with safe sockets that you can use with confidence.
- Light and heat (or a fan in the summer) provided by self-same electricity supply.
- A double hotplate (one for experiments and one for you).
- A refrigerator for keeping ingredients cold and providing a steady supply of ice.
- Cool, neutral colours.

② EQUIPMENT AND STORES

Along with your shed, you'll need some basic equipment to make the lab run smoothly, as well as the 'ingredients' for your experiments. Since you'll be using some of these more often than others, it makes sense to start collecting them as you go along, so that you're ready when the urge to make a discovery comes calling.

The experiments have been put together so that many of the items in your stores cupboard are used repeatedly. In addition, if you look over to the basic stores checklist you'll see that it includes the kind of things that you're likely to have lying around the house or the shed anyway. This is part of a happy design that's intended to make it easy to start the experiments in the book and also to demonstrate that science – and everything you need to explore it – is all around us.

Whatever you do, don't buy any of this stuff new unless you have to. One of the joys of shed science is picking up the equipment you need second-hand or in remainder shops. Re-purposing old junk that other people want to get rid of is not only good economic sense but promotes sustainable shed science as well. Hold your head high and get it on the cheap!

Finally, a word on buying chemicals. It's not as easy as it used to be, but there are various sources you can try. Your local pharmacist can help with many items; if not, try hardware stores, DIY and garden centres, and photographic equipment suppliers. And don't forget the good old-fashioned chemistry set, which – together with refills – usually contains a range of interesting and useful chemicals. As always, the Internet is a great way to track down those harder-to-get-hold-of items.

Storing Chemicals Safely

The best place to store any chemicals is in a cool, lockable cupboard. There are also some basic tips to follow. First, always store according to the manufacturer's instructions, and store incompatible chemicals separately. Don't expose any chemicals to direct sunlight or heat, and make sure that the containers are sound and labelled properly.

BASIC STORES

For the experiments contained in this book, the following basic items will be necessary:

- Water and ice
- Vinegar (various)
- Baking soda
- Food colouring, cinnamon
- Salt, sugar, bread, rice, crackers, eggs
- Coloured ink
- Cooking oil
- Washing-up liquid
- Various fruits including grapes, lemons, apples, pears
- Various pickles
- Various vegetables including potatoes, broccoli, carrots, red cabbage
- Tea bags, gin, beer

BASIC EQUIPMENT

The serious hardware has been covered in The Shed Laboratory on pages 8 and 9, but there are some everyday items that you should have available to make a wide range of experiments work:

- Goggles, gloves, oven gloves
- Old saucepans, washing-up bowl, glassware of various sizes, bucket and plant pot
- Clean empty tin cans, empty plastic drinks bottles, plastic containers, old aluminium dishes, styrofoam plates, cardboard boxes, odd bits and pieces of wood, sealable sandwich bags
- Sieve
- Mirrors, magnifying glass
- Camera
- Batteries of various sizes
- Magnets
- Drinking straws, lollipop sticks, toothpicks
- Sticky tape, plasters and glue
- String, cotton thread, elastic, wool
- Corks and modelling clay
- Paper, blotting paper, tracing paper, stiff card, cardboard, newspaper, toilet paper
  - Pencil and sharpener, biros
  - Ruler, protractor, drawing compass
  - Scissors, tin opener, tweezers
  - Spoons
- Craft knife
- Nails, tacks, screws, nuts, bolts of various sizes
- Pins, needles, and drawing pins.
- Paper clips
- Matches, tapers, candles
- Silver foil, sandpaper, steel wool, cotton wool
- Balloons
- Bicycle pump, football adaptor
- Simple general tools – drill, saw, screwdriver, hammer, pliers, tongs, jigsaw, staple gun

Should you spill anything, clean it up straight away, observing the manufacturer's safety procedures. Finally, when using chemicals, make sure there's adequate ventilation, handle them carefully, and observe any specific manufacturer guidelines.

SPECIALIST STORES AND EQUIPMENT

Of course, there'll be times when you need to get hold of something a bit specialised – usually for an experiment involving chemistry or electricity. For these experiments, you'll need some of the items listed below.

Chemicals

- Sodium hydroxide
- Sodium Bisulphate
- Sulphur
- Hydrated lime
- Sulphate of Ammonia
- Camphor
- Meat tenderiser
- Tincture of iodine
- Starch powder
- Borax
- Potassium chloride
- Surgical spirit
- Cream of tartar
- Isopropyl alcohol
- Aluminum sulphate
- Potassium ferricyanide
- Copper acetate monohydrate
- Sodium silicate solution

ADVANCED EQUIPMENT

Some of the experiments need specialised equipment, but it's all easy to track down. Electrical stores, mail order and the Internet are your best bet for most components. Alternatively, you'll be surprised at what you can pick up at car-boot sales and junk shops.

- Large pizza box
- Neodymium iron boron (NIB) magnets
- 35mm film canister
- Various kinds of electrical wire
- Platinum-coated wire
    - Wire cutters
        - Multimeter
    - Battery clip
    - LEDs (light-emitting diodes)
        - Ceramic light fittings
        - Copper flashing
    - Alligator clips
    - Test tubes
    - Lab stands
- Ring clamps
- Variable transformer
- Extension lead
- Blender
- Medicine dropper
- Soft copper tubing
- Large mirror
- Germanium diode
- Piezoelectric earphone
- Funnel
- Microscope
- Large piece of plywood
- Larger piece of plastic sheeting
- Fluorescent bulb
- Steel ball bearings

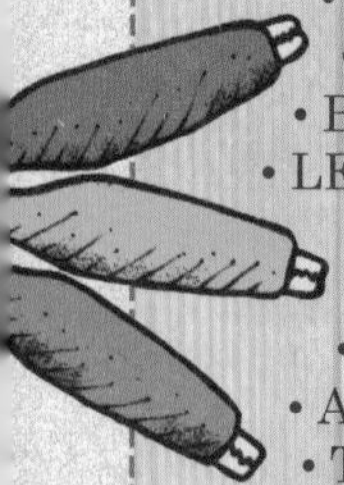

③ SCIENCE PRIMER

Archimedes had the right idea. 'Give me a lever long enough,' he said, 'and a fulcrum on which to place it, and I shall move the world.' Now there was a man with a plan – a man who understood the power of science.

These days, statistics tells us that science has fallen out of favour, both as a subject to study and as a career to pursue. Since there's very little in the modern world that owes no debt to science, this is something of a puzzle. Every time you make a cup of tea or turn on the television, run a bath, load up the iPod, put on a pair of socks, mow the lawn, order a pizza or use your toilet, you're taking advantage of one scientific discovery or another, so why should we be so indifferent to something that touches all of us? How did we find ourselves in an age that loves gadgets but loathes science?

This book hopes to play a small part in righting that wrong. It hopes to inspire men, women and – under proper supervision – kids to rediscover the joys of discovery; to find out what it feels like to prove that a theory holds water, that something can be demonstrated in the real world rather than just theorised in the pages of a book.

Now, because science has been relegated to the back of the class, shed scientists may have some ground to make up, so here's a quick primer to get you on the right track.

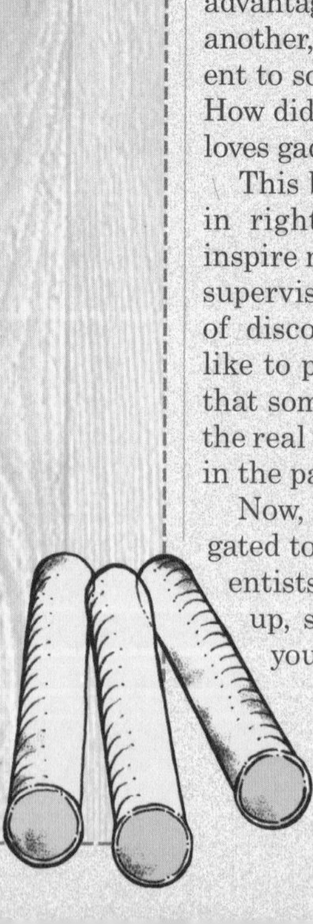

Fig. 4. *Test tubes are one of the few 'proper' items of scientific equipment you'll need.*

Analysis and Evidence

Science isn't really about biology, physics and chemistry (or any of the more popular modern additions like ecology and life science); it's really a way of making sense of the world in an organised fashion, using evidence that you can measure. That means your most important scientific tool isn't anything you can buy in the shop – it's your brain and how you use it.

Once you've got your head around that, then the actual scientific method is probably less complicated than you think. You can try breaking it down like this:

- You observe something that's going on around you.
- You then attempt to come up with a way of describing that object or event; this is usually called a hypothesis.
- You use that hypothesis to make predictions about related objects or events.
- You test your hypothesis by creating experiments that allow you to make new observations and adapt your hypothesis depending on how the experiments turn out.
- Once the hypothesis is proven over a period of time, it becomes a theory.

Conducting Experiments

One of the most fantastic things about science is that you don't have to believe what anyone else says. There are no leaps of faith required. If you want to verify whether something that someone else says is true, you can simply redo the experiment according to their instructions. If it works, then it's true. If it doesn't…

Of course that means you need to be rigorous. You need to follow instructions carefully, use the correct ingredients, do things in the right order and allow the right amount of time for reactions to take place.

Finally, if you're serious about science, then it's also a good idea – and good practice – to take notes as you go through an experiment and then write it up when you've finished. That way you can assemble a home library of experiments that you can pass on to your kids so that they can follow in your footsteps. As Newton pointed out, it's not like we're going to run out of things to discover:

'I do not know what I may appear to the world, but to myself I seem to have been only like a boy playing on the seashore, and diverting myself in now and then finding of a smoother pebble or a prettier shell than ordinary, whilst the great ocean of truth lay all undiscovered before me.'

Fig. 5. *Conducting experiments is central to proving scientific theory – what's more, it's something that everyone can enjoy.*

4 SAFETY FIRST

Any journey of discovery involves an element of risk, but by adhering to the simple rules on the following pages you can make sure that the risk – whatever it is – doesn't overtake you.

Necessarily, this is going to come over as a long list that starts 'Don't do this,' but really, with a very few exceptions, it's all common sense. The exceptions are the electrical and fire experiments, which warrant their own special heading. In the meantime, remember that the whole purpose of this book is to help you understand more about the way things work, generally using simple household items that everyone can get hold of, in order to take some of the mystery out of science and put some of the fun back in.

Fig. 6. *Goggles and gloves are essential lab accessories for those experiments that involve chemicals and chemical reactions. They look good, too.*

Safety in Snapshot

If your shed is calling and you're keen to get started on some of the experiments in the book, then just remember these simple rules:

- None of these experiments or the ingredients they contain is edible – even if they look tasty. (The exception of course is Tickle Your Taste Buds on page 56, which won't work unless you eat it.)
- Make sure that you wear sensible clothing all the time; if you have long hair, tie it back or wear a hat.
- Wear safety clothing including goggles and gloves when you're handling chemicals or hot objects.
- Any ingredients should be stored out of the way and according to the manufacturer's instructions.
- Read through each experiment first and make sure you understand what's required before you start.
- Some of these experiments fly – never point them at anyone.
- When you're working with any of the ingredients, keep your hands away from your face and eyes; wash up thoroughly afterwards.
- For experiments that require FIRE or ELECTRICITY – see the appropriate headings below.
- Make sure you clean and tidy your lab after each experiment and return any ingredients to their proper place.
- When you're not around, make sure your shed is secured and padlocked in case your kids or the neighbours' kids come calling. Kids are genetically programmed to be drawn towards shed science but similarly programmed to be unaware of the potential dangers.

General Safety Advice

Don't try these experiments if you're tired, or on medication, or while enjoying a cold beer. Take your time with the experiments – you'll stand more chance of completing them successfully and, after all, there's not exactly a deadline for any of this. Don't substitute any of the ingredients here with something else unless you know what you're doing, and if you've got any doubts about anything here – especially those experiments that involve electricity or fire – then DON'T DO IT.

Other tips are more general. Always try and work in good light so you can see what you're doing. If your shed doesn't have electricity, then experiment during the day when you're less likely to make mistakes or drop things. Similarly, ventilation is important. Consider working with the door open or at least cracking a window to get a good flow or air in and out of the lab. Very few of the chemicals used in this book are actually noxious but lots of the ingredients will give off a good pong – especially in a relatively confined space.

Electrical Experiments

Although very few of the experiments in this book need mains electricity, those that do require you to take extra care and preparation.

- Always use electrical appliances in accordance with the manufacturer's instructions.
- If you're fiddling with an appliance, make sure it's switched off and unplugged.
- Water loves to conduct electricity, so never touch water that has current flowing through it and keep

Fig. 7. *Electricity only plays a small role in these experiments, but it's nevertheless an important one. Even more important is that you use it carefully and sensibly.*

electrical appliances and water away from each other. Make sure your hands are dry as well.
- Remember that because you're mostly made of water, you conduct electricity really, really well.
- Do not attempt the Glow-in-the-Dark Gherkins Experiment (pages 58 and 59) unless you have access to a variable transformer or the equivalent.
- An electric shock may not kill you but it'll seriously hurt.
- If you suspect that someone has been electrocuted, don't touch them. Switch off and unplug the source of the shock at the mains. If you can't, try to separate them from whatever gave them the shock using something that doesn't conduct electricity, like a piece of wood or a rolled-up newspaper. Check to see if they're conscious; if not, start CPR. Call for help as soon as possible.

15

IMPORTANT SAFETY NOTE

For those experiments where a little extra caution or some parental supervision is required, we've added the little symbol shown top left. Such experiments involve matches, chemicals and other potentially hazardous elements, and so should always be conducted with care and safety.

Experiments with Fire

Always treat fire with the kind of respect you'd give to anything that can burn you. Kids should never be allowed to use matches, tapers or any other kind of fire or flammable liquid without supervision. Keep the items in the Shed Safety Kit to hand, and if someone is burned, start by cooling the affected area under running water for at least 10 minutes. Don't try and ease the pain with creams, don't use sticky plasters, and leave any blisters well alone. A clean plastic bag makes a good emergency dressing. Remember that those experiments here that use solar power are going to get hot as well.

Shed Safety Kit

It's a good idea to have the following items in your shed or in your home (should you have to transfer some of the experiments to there):

• First aid kit
• Fire extinguisher
  (a bucket of sand is a useful alternative in the shed)
• Fire blanket

Warning Symbols

There are various internationally recognised safety symbols that you might see used on packaging; for reference purposes, here are some of them and what they mean:

You must wear eye protection, such as a pair of goggles

Highly flammable – e.g. alcohol or surgical spirit

Irritant/Hazard – e.g. borax or potassium chloride

Corrosive – e.g. sodium silicate or sodium hydroxide

Toxic – e.g. iodine or copper chloride (neither of which are used in this book)

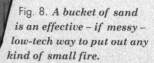

Fig. 8. *A bucket of sand is an effective – if messy – low-tech way to put out any kind of small fire.*

5 HOW TO SPLIT WATER

Water is simple stuff. It's made up of two gases (hydrogen and oxygen) and with the right equipment you can 'split' it to reveal these two chemical components. This easy experiment shows you how.

What You Need
The equipment needed to split water into hydrogen and oxygen isn't as complicated as you might think. All you need is two ordinary HB pencils, a pencil sharpener, some salt, a piece of thin cardboard, some lengths of electrical wire, a glass, some warm water and a 9-volt battery.

What To Do
Sharpen both ends of the two pencils, and then cut the cardboard to an oblong shape (this should sit comfortably on top of the glass, but stick out slightly at either end). Make a couple of holes in the cardboard, either with a pair of scissors or by pushing the pencils through – the pencils need to fit snugly so that they'll stay put when you push them through. Get a level teaspoon of salt and dissolve it into the warm water. Take a length of the wire and wrap it round the lead of one of the pencils at the top, then attach the other end to the positive terminal of the battery. Attach the second wire to the second pencil lead and wrap the other end round the negative terminal. Lower the other end of your cardboard-and-pencil contraption into the water. As the current flows, you'll see bubbles collecting around the bottom points of each pencil.

How It Works
The physics behind this experiment is deceptively easy. The pencils act as electrodes, and when the electricity from the battery passes between them the water is split into its component parts. The battery isn't powerful enough to split water on its own and so it needs help from the salt, which is a good conductor. However, by adding salt you've added an extra element to the process: sodium chloride. In this reaction, chlorine gas from the salt gathers around one pencil tip and hydrogen around the other. If a higher current is passed through the solution, then the water does split into hydrogen and oxygen and the process is called electrolysis.

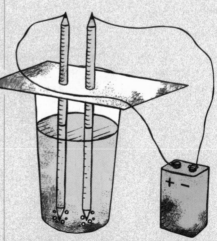

Chlorine gas gathers around one pencil tip...

...and hydrogen gas around the other

6 MAKING SLIME

Boys of all ages like a good dollop of slime. You can stick your hands in it, throw it around and leave it lying about as a nasty surprise for people to find, and it makes a great decoration at Halloween. Fortunately, making your own slime is a very straightforward task that requires just a few easy-to-find household items. This experiment shows you how, with simple step-by-step instructions.

What You Need

Some white glue (a non-washable type), sodium tetraborate (available as borax from your local supermarket – try looking in the laundry-detergent section), water, some sealable plastic bags, a stirrer such as a spoon or craft stick, a jar or large cup, and a bowl to mix all the ingredients.

What To Do

Place an equal quantity of the glue and water in a bowl. Mix well with a spoon or craft stick.

Using a separate container, combine a tablespoon of sodium tetraborate powder with a cup or so of water, and stir. It's easier if you use a jar with a lid, as you can then screw on the lid and shake the mixture well. If all of the sodium tetraborate powder dissolves, you need to add a bit more. When you get to the point where no more sodium tetraborate will dissolve, the solution is saturated.

Now add about two tablespoons of the sodium tetraborate solution to the bowl with the glue-and-water mixture, and stir quickly. The resulting mixture should be slimy or gooey.

Knowing just how much solution to add is the trick to this experiment. If you add too little, your slime will contain excess glue and it will be sticky; if you add too much, your slime will be very wet. Touch your slime with your

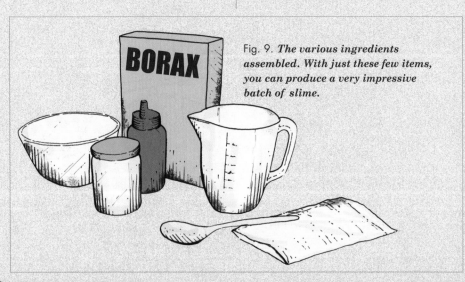

Fig. 9. *The various ingredients assembled. With just these few items, you can produce a very impressive batch of slime.*

hands when it doesn't look like a liquid anymore. If your slime feels sticky, try adding a little more solution. If your slime feels very wet and slippery (but is not still runny), remove it from the container and knead it in your hands. In a few minutes, any extra sodium tetraborate solution will evaporate or be absorbed.

You can save your slime for a long time by putting the stuff into sealable plastic bags. If your slime dries out, you can add a bit of water back into it. If it gets too dry, you'll have to start again. Also, to colour your slime, use food colouring.

Fig. 10. *Using a separate container, combine a tablespoon of sodium tetraborate powder with a cup or so of water, and stir. It's easier if you use a jar with a lid, as you can then screw on the lid and shake the mixture well.*

How It Works

This experiment demonstrates how polymers are made, and how other chemicals affect them. Polymers are used in nearly everything – for example, most kinds of plastics, nylon, and clothes – and these can often be spotted by their name: if it ends in '-on', like nylon or rayon, it's probably a polymer. When you mix glue with a bit of water, you make the polymer, polyvinyl acetate (PVA). The sodium tetraborate solution is a 'cross-linking' substance that binds the polymer chains together to make the glue solution thicker. As the polymer chains get more 'bound together', it gets harder for them to move around, and your slime starts to be more like putty.

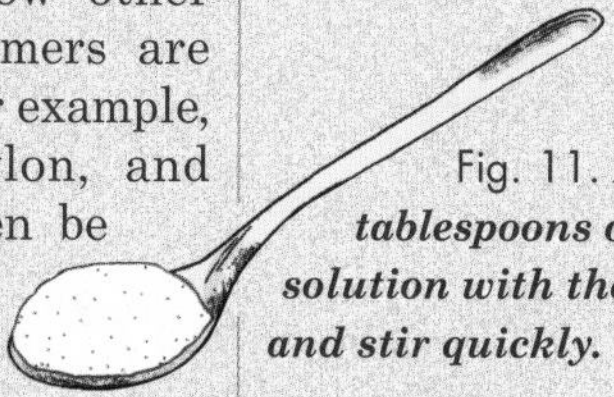

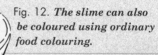

Fig. 11. *Now, add about two tablespoons of the sodium tetraborate solution with the glue-and-water mixture, and stir quickly.*

Fig. 12. *The slime can also be coloured using ordinary food colouring.*

⑦ How to Pick Up Ice

It's incredibly hard to pick up an ice cube, especially one that's started to thaw. However, it can be done thanks to this simple science trick. You'll need an ice cube, a match, some ordinary table salt and a bowl of water.

What To Do

Fill the bowl with tap water (it doesn't matter whether you let it run cold or not) and then pop the ice into the water. Gently place the match across the ice cube and then sprinkle salt around it. Wait for 20 minutes, hold the match ends between your finger and thumb, and then pick up the ice cube using the match. Simple.

How It Works

The salt melts the ice around the match but, because it can't get underneath, the match freezes to the ice cube. In addition, the action of melting pulls heat from the surrounding water, so while the rest of the cube is melting, the match actually freezes to the top part. Remember that fresh water freezes at a higher temperature than salt water.

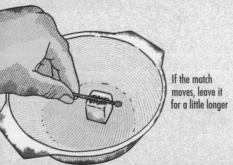

If the match moves, leave it for a little longer

⑧ Perpetual Motion

OK, so it isn't really perpetual motion (that's Nobel Prize material), but this simple trick will run for days and days. The equipment is straight forward: a cork, four needles, a bowl, some water and four small pieces of camphor.

What To Do

Cut a slice through the cork so you get a slim cork disk. Cut four small squares of cork at the same time. Push the needles into the cork disk to make a cross with the cork at the centre. Cut a tiny notch out of each of the four cork squares and pinch a piece of camphor into each one. Then push a cork square onto each end of the needles – with the camphor always on the same side.

Lower the contraption into the water and wait. After a while it will begin to spin, and it can keep moving for days.

How It Works

As the camphor dissolves it forms a solution with the water. Since this has a lower surface tension than ordinary water, your cork can spin freely in the bowl. (If you shave bits of camphor into the bowl, they'll dart around all over the place because of the different surface tension at the edges.)

9 THE REPULSIVE GRAPES

In this experiment you'll discover that the humble grape doesn't feel at home near the South or the North Pole. The question is – why?

What You Need

A neodymium iron boron (NIB) magnet; this is sometimes called a 'neodymium' or 'rare earth' magnet, and is more powerful than a traditional magnet. You'll also need a couple of similarly sized grapes, a straw, a 35mm film canister (with lid), a ruler and a drawing pin.

What To Do

Use the ruler to find the middle of the drinking straw and then use the drawing pin to make a small hole there. Wiggle the pin about a bit to enlarge the hole slightly – it needs to be large enough for the finished device to spin round freely. Next, take the pin and push it from the inside through the canister lid, so that it's sticking up through the top. Put the lid back on the canister.

This is the least scientific bit: you need to push a grape onto each end of the straw and position them so that everything balances when you lower the hole in the straw over the drawing pin. You may have to adjust the grapes a bit to achieve this but, hey, that's why it's called experimenting. Move the magnet so that one of its poles is facing one of the grapes, and watch as it pushes the grape away. Turn the magnet round – the same thing will happen with the other pole.

How It Works

Ferromagnetic material (for example, iron) is strongly attracted by both magnetic poles; paramagnetic material (like aluminium) is weakly attracted to both poles. Water – which makes up most of the grape – is diamagnetic, which means that it's repelled by both poles of a magnet. As a force, diamagnetism is pretty puny, which is why you need a very strong magnet and something watery and light like a grape in order to see it in action.

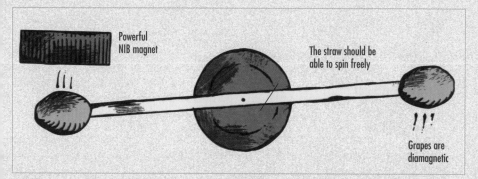

Fig. 13. *The water content of grapes makes them diamagnetic, which means they're repelled by both poles of this very strong neodymium iron boron (NIB) magnet.*

⑩ WATER INTO WINE

This is an absolute classic that, with practice, will work every time. However, you need time to prepare the trick and you must insist that no one tastes the 'wine', because that'll definitely give the game away (nor is it safe to do so).

What You Need
There are no funny prop glasses with hidden compartments – this is pure science at work. You'll need three small glasses; it's more effective visually if the one that contains the 'wine' is an actual wine glass, but it doesn't matter too much. You'll also need a few drops of sodium hydroxide, a few drops of phenolphthalein, some concentrated acid (something like sodium bisulphate), and some water.

What To Do
Start by pouring water into the first glass, and then add a couple of drops of the sodium hydroxide solution. Next, add a few drops of phenolphthalein to the empty wine glass, and put a few drops of the concentrated acid into the third glass. Let them all sit there while your audience files in.

When you're ready, pour the clear water into the wine glass. As you will see, it turns red. Show your audience.

Water into wine. Whatever you do, don't let anyone drink it!

Acid or Alkaline?
This experiment is to do with the pH scale, which is used to express the concentration of hydrogen ions (that is, positively charged atoms of hydrogen) in a solution. A low concentration of hydrogen ions corresponds to a high pH, while a high concentration results in a low pH. Anything that you add to water that raises the pH is alkaline (it's also often called a 'base') and anything that lowers pH is an acid. The changing colours are the result of changing pH levels in the solution. See The Cabbage Chameleon on page 92 for more information on pH scales and indicators.

REAL SCIENCE

Take the applause. Then, to further astound them, pour it from the wine glass into the third glass and it will turn clear again.

How It Works
The sodium hydroxide makes the water in the first glass alkaline. When you pour the alkaline solution onto the phenolphthalein, a chemical reaction occurs, turning it deep red. When you pour the red mixture into the glass with the concentrated acid in it, the phenolphthalein returns to a colourless form – water into wine, and then back again.

11 EGG AND BOTTLE TRICK

When you explain to your friends that you're going to squeeze an egg into a bottle through an opening that's smaller than the egg itself, they're going to think you're having them on. Get them to bet against it if you like – they're going to be the ones with egg on their face (or rather not, because it's going to be in the bottle).

What You Need
Any glass bottle with a neck that's slightly smaller than the circumference of the egg, a box of safety matches, and a smallish cooked egg. For safety, boil two eggs in case the first one gets mashed up.

What To Do
Boil your eggs for about ten minutes and then set them aside until they've cooled sufficiently for you to shell them without turning into an egg juggler. Put the egg in the neck of the bottle (the narrower end should be pointing down) and let it sit there. This is a really boring part of the experiment because nothing's going to happen no matter how long you wait or how many observations you make on your clipboard. So, take the egg out, light three or four matches in turn, and drop them into the bottle. Then put the egg quickly back into the top. As the matches burn, the egg will be 'sucked' magically into the bottle.

How It Works
The invisible force that pushes the egg into the bottle is air pressure, of course. At first, with the air inside and outside the bottle the same, the egg just sits there because it's being neither pushed nor pulled. However, burning matches consume oxygen, decreasing the air pressure inside the bottle. The greater air pressure outside thus pushes the egg down into the bottle – or it just gets completely stuck and you'll have to dig it out and try again with more matches and the second egg.

Carefully drop three or four lit matches into the bottle

Watch as the egg is 'pulled' into the bottle as if by magic

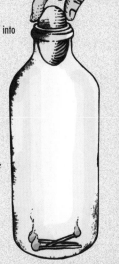

Fig. 14. *As the matches use up the oxygen in the bottle, the egg will be sucked in through the narrow neck.*

12 NATURE'S BATTERIES

Is it possible that there's more going on inside your favourite fruit and vegetables than you'd normally think? That far from just being tasty – and usually healthy – they're also able to pack an electrical punch that's strong enough to see and feel? Let's find out.

What You Need

If you just want to set up the basic 'battery', all you need is a lemon, some copper wire, a paperclip, some wire cutters and some sandpaper. If you want to try something more sophisticated – and powerful – substitute some zinc for the paperclip (it's a better conductor) and add more lemons and more wire for the terminals. You'll also need something to measure the electricity (such as a multimeter), or a small electronic device like an LED (light-emitting diode) to prove that the batteries are working.

Copper wire (left) and steel wire (right)

What To Do

For the basic experiment, strip the insulation from the copper wire until you can cut about 5cm (2in) of bare wire off. Bend open the paperclip and cut it to the same length as the copper. Sandpaper off any rough spots on both lengths of metal. Roll the lemon on the table to loosen the insides. Make sure you don't break the skin. Next, push both lengths of metal into the lemon. They need to be as close together as possible without actually touching. Wet your tongue and touch both ends of the metal at the same time. You'll feel a tiny tingle and will probably have a metal taste in your mouth.

A single lemon will produce around three quarters of a volt, which isn't really enough to power much of anything. If you want to crank up the current, you'll need a larger supply of lemons. You'll also want to increase the area of your copper and zinc elements (the bits that go into the fruit) by flattening them with a hammer. Simply 'chain' the lemons together using electrical wire and going from the '+' on one lemon to the '−' on the next and so on. Hook four lemons up like this and then connect the loose ends to the LED. If you look closely at the base, you'll be able to see that although it's almost circular there is a flat section where one of the wires comes out. Connect that wire to the '−' side of your lemon array. Connect the other wire to the '+' side and voila! – there is light.

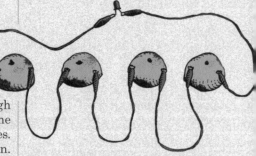

Fig. 15. *By chaining lemons like this you can boost the amount of electricity they generate significantly so that it's enough to light a tiny LED.*

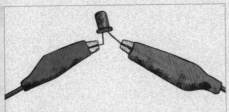

Fig. 16. *When you complete the circuit the LED will light up.*

delivered. So, even if your lemon array is producing a decent voltage, the resistance in the circuit may impede the amount of current passing through it.

Even one lemon can generate electricity

How It Works

Fruits and vegetables like lemons and potatoes act like the kind of batteries you buy in a shop. In both, chemical energy is the power source and the results are electrons that collect on the negative terminal of the 'battery' and then flow to the positive terminal as fast as they can – once the circuit has been completed. When you stick your moistened tongue on both terminals/electrodes, you're completing the circuit, which is why you feel a tingle. Meanwhile, inside the lemon, the atoms in the acid are busy grabbing electrons from the copper atoms and shooting them across to the steel (or zinc) atoms.

Why don't we use fruit for batteries all the time? Well, apart from the fact that they've got a limited lifetime (and the smell), they're not very efficient. That's because something that generates a reasonable voltage may still not be able to 'drive' even a simple electronic device like an LED because it's not producing enough current. Yes, there's a difference. The easiest way to think of it is to imagine a garden hose with a twist in it. When you turn on the tap, there's plenty of good water pressure until it hits the twist – after that, the pressure diminishes. Voltage is a bit like the water pressure, while current describes how efficiently that voltage is

Battery Power

Alessandro Volta (1745–1827) was an Italian physicist from Lombardy who is remembered as the father of the modern battery. His early experiments involved wine cups filled with salty water into which he placed electrodes made of various metals. Eventually he hit upon zinc and silver as being most effective. This work formed the basis of the so-called voltaic pile, the first modern electric battery, which he built in 1800. Volta also gives his name to the volt – the internationally recognised way of measuring the strength of an electrical source.

REAL SCIENCE

13 THE SNAKE CHARMER

This is an experiment that uses static electricity and – thankfully – doesn't involve a real snake. It does, however, produce an unexpected special effect that will have your friends scratching their heads. You'll need some tissue paper, a pair of scissors, a plastic biro pen, something made of wool, and either a lid made of tin or the kind of tin plate that you take camping.

What To Do

Use the scissors to cut a spiral shape out of the tissue paper; it needs to be wide enough so that when you hold up the end it looks like a snake preparing to strike. Draw a face on the head of the snake and then place it on the tin lid/plate. Bend the head of the snake very slightly up towards you. Rub the plastic pen vigorously up and down on the wool for a minute, then hold the pen near the 'head' of your snake and wave it about. The snake will rise to meet the pen as if charmed.

How It Works

When you rub the pen against the wool you give it an electrical charge, which attracts the paper snake. Because the paper is so light, the attraction is enough to lift it off the metal plate.

14 THE WATCH COMPASS

If you have a wristwatch – or even if you haven't – you can always tell where North and South are, thanks to this ingenious invisible compass. You'll need an analogue wristwatch. If you have a digital one, you can still picture where the numbers are in your mind's eye and the compass will still work.

What To Do

If you're in the northern hemisphere, point the hour hand at the sun. Due south will be exactly at the mid-point between the hour hand and 12 o'clock. (North is obviously the opposite.) In the southern hemisphere, it's slightly different. You have to point the 12 on your watch at the sun and then imagine a point between that and the hour hand. That will give you due north.

In the southern hemisphere, line 12 o'clock up with the sun

How It Works

Because the sun always rises in the east and sets in the west, you can use your watch (or your imagined watch) to help you work out where north and south are.

⑮ TEST YOUR REACTIONS

The idea that the human body is actually controlled by tiny people living in your head, looking out through your eyes and pulling levers to make you do things, has often been explored in films, cartoons and comic books. But just how quickly can your brain get a message to your muscles? And would you actually be any faster if there really were little people pulling the strings and controlling your every move? Find out with this simple experiment.

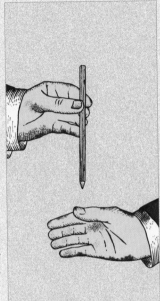

What You Need
A pencil and a friend. You can try and do it by yourself, but you'll usually win, for reasons that are explained in the How It Works section.

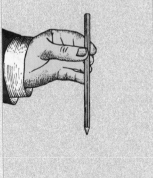

What To Do
Tell your friend to stand opposite you with their hand held out as if they were reaching for a glass of water. Hold the pencil above their open hand and then lower it, so that it's suspended inside their hand. Let go of the pencil and see if they can catch it. This is much harder than it looks, and they'll miss the

pencil nearly every time. If their reactions are particularly good, make it more difficult by getting them to catch the pencil between their forefinger and thumb.

How It Works
When you let go of the pencil, your friend sees it begin to fall. His eyes send a message to his brain that the pencil is falling, and the brain then responds by sending a message to the open hand to close and grab the pencil before it falls to the ground. All of this only takes a split second, but it's usually long enough to miss the pencil. If you do it to yourself, however, you'll be able to catch it nearly every time because your brain already knows when you're going to let go of the pencil and so anticipates the drop and has time to grab it before it falls to the ground.

Fig. 17. *If it's too easy for your partner, get them to try to catch the pencil between their forefinger and thumb instead. They'll find that much more of a challenge.*

16 THE TWO-TONE FLOWER

This marvellous experiment not only clearly demonstrates a sound scientific principle, but also results in something rather lovely. When science and nature combine to produce such beauty, we can only stand back and admire.

What You Need
Two ordinary glasses, some tap water, two different-coloured inks (red and green or red and blue will work well), a sharp knife and any large-headed white flower, such as a carnation.

What To Do
Fill each glass to about three-quarters full with the water. Add a few drops of one ink to the first glass and then a few drops of the other ink to the second glass. Swirl them round to mix in the ink. Take a knife and carefully split the stem of the flower to about halfway up, then place one half of the stem in one glass and the other half in the other glass. In an hour or two, you'll see that the white head of the flower is now two-toned – each half having taken on the colour of one of the inks in the glass. It's an amazing effect.

How It Works
It's all down to two things: capillary action and transpiration. Capillary action is the ability of a narrow tube to draw water upwards, seemingly against the laws of gravity. This is

Adhesion and Cohesion
Water molecules are attracted to each other – if you will, they like to stick together. This is because the oxygen part of water is negatively charged and the hydrogen bit is positively charged. The hydrogen atoms of each water molecule are attracted to the oxygen atoms of the water molecule next to it, and this is called 'cohesion'. When water molecules are attracted to something else – like the inner surface of the plant stem – this is called 'adhesion'. Cohesion and adhesion are key to the way plants use capillary action to draw water upwards from their roots, against gravity, to the parts where it is needed.

possible because the molecules that go to make up water like to stick to things (including other water molecules), which allows the stem of the plant to pull the water up towards the leaves and – in this case – petals. The colour is so strong because the dye remains in the flower head while the water disappears into the atmosphere through tiny holes in the plant – this is transpiration.

17 THE STINKTASTIC BOMB

Apparently, the U.S. military is busy conducting research into a battlefield stink bomb with an odour so vile that it renders enemy soldiers unable to fight. While it's unlikely that this experiment will provoke quite such an extreme reaction, it's certainly one of the most disgusting smells imaginable.

What You Need
All of these ingredients are unpleasant enough on their own, so take care and refer to the Safety First section of this book on pages 14–16 before you start. You'll need some sulphur, some hydrated lime and some sulphate of ammonia. These are available from garden centres and DIY stores that carry basic building supplies. You'll also need something to put the 'bomb' in, an old stove pot, a bucket for mixing and some clingfilm.

What To Do
Take 110g (4oz) of sulphur and 220g (8oz) of hydrated lime and mix them together in an old pot – preferably one that you don't need any more. Add a litre of water and heat on top of the cooker until everything's mixed nicely. You'll end up with a yellow liquid, plus a mess of lime at the bottom of the pot. Pour the liquid into a bucket, carefully take it outside and add a pound of sulphate of ammonia. Stir with an old stick, then cover with clingfilm and leave it for about 30 minutes. Pour the mixture through a cloth into a bottle (or whatever you're using to store the stuff). You can then decant small amounts into smaller bottles as you need it – a little goes a long way.

How It Works
The active ingredient is ammonium sulphide, which is produced by mixing the three constituents together. This gives off that familiar rotten-eggs stench and is both powerful and long-lasting. Use with care.

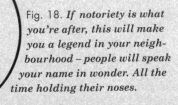

Fig. 18. *If notoriety is what you're after, this will make you a legend in your neighbourhood – people will speak your name in wonder. All the time holding their noses.*

18 CREATE A CHEMICAL GARDEN

No green fingers, or worse, no garden to speak of? It doesn't matter. This is the twenty-first century, and you can create landscapes that defy the imagination and can be cared for as if they were real gardens. Of all the experiments included in this book, this is one of the most extraordinary.

What You Need

To start off, you should grow some salt crystals—it's easy and it will help you to understand the processes involved in this project. For that, all you need is a small glass jar, a pencil, some cotton thread and a packet of household salt.

The requirements are more complex if you want to grow an actual garden, but they're still easy enough to get hold of. It's important to understand that a chemical garden grows in three stages. First, you need the basic chemical components; second, these must be turned into the crystal 'seeds' to 'plant' in the garden; and, third, you need to provide the right environment for the garden to grow.

To grow the crystals themselves, try getting some aluminium sulphate (sometimes known as alum) potassium ferricyanide (sometimes called ferri) and copper acetate monohydrate. See A Chemical Bestiary on pages 10 and 11 for details of where you can buy chemicals and how to handle them properly. You'll also find important hints and tips in the Safety First section (pages 14–16). In terms of hardware,

you'll need water, a saucepan and something to heat it on, a pair of tweezers, and a saucer.

For the garden itself, you'll need a large glass container. Any shape will do, though an old fishbowl always looks nice. To go inside it, get enough sand to cover the bottom to a depth of about 6mm (1/4in) and enough sodium silicate solution to half fill the container – you'll top the rest up with ordinary water.

What To Do

To grow salt crystals simply and quickly, try this. Take the small jar and fill it halfway up with warm water. Stir salt into the water until you can't dissolve any more. Tie the thread onto the pencil around its middle, and lower the loose end into the jar. Make sure it doesn't touch the bottom; if it looks like it will, just wind the pencil round a few times and fix it with sticky tape. Let the water evaporate and, as it does, you'll see cubic salt crystals forming all along the thread. Simple as that.

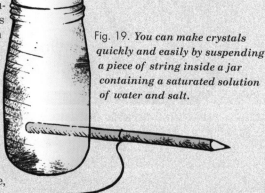

Fig. 19. *You can make crystals quickly and easily by suspending a piece of string inside a jar containing a saturated solution of water and salt.*

Salt crystals 'growing'
on a piece of string

If you want to go the whole hog and grow crystals for your garden, here's how you do it. We'll use alum as the example. Put a pint of water into the saucepan and add about 100g (4oz) of alum. Heat the mixture gently, stirring all the time to make sure the alum powder dissolves. As you did with the salt in the simple version of this, keep adding alum until it stops dissolving, then turn off the heat and let the saucepan cool down. When it does, pour some of the mixture into a saucer and leave that somewhere to cool. After a couple of days the liquid in the saucer will evaporate, leaving the crystals behind. Choose a few of the big ones as your 'seeds'.

You can make crystals using potassium ferricyanide and copper acetate monohydrate in exactly the same way, though the amounts required vary slightly. Use 100g (4oz) of potassium ferricyanide to every 200ml (third of a pint) of warm water, and use 20g (3/4oz) of copper acetate monohydrate in 200ml (third of a pint) of hot water.

OK, so you've got your crystal seeds, now you need to set up your garden. Take your large glass container and pour in the sand. After that, half-fill it with sodium silicate solution and then top it up with ordinary water. Using the tweezers, carefully drop in two or three crystal seeds from each chemical, taking care that the seeds aren't too close

to each other. You have now planted your garden. Expect the crystals to grow to up to 12cm (5in) over the course of a few days. When they've stopped growing, with care you can replace the solution with ordinary water.

Incidentally, rather than making your own, you can buy kits that contain all you need to make a simple chemical

Fig. 20. *As the water evaporates, it leaves behind crystals that are used as 'seeds' for the chemical garden.*

Silent Running

Back in 1984 a group of students from the south of England won a competition to have an experiment conducted in outer space on the Space Shuttle. Their idea? A chemical garden. The students proposal was to discover what effect microgravity would have on the symmetry and direction that the chemical plants grew in. To convince the judges they even developed a computer simulation to predict what would happen. In space, the results in the real garden were amazing. Many grew in random shapes with spectacular contortions, while others grew in almost perfect spirals. As yet, the differences in shape and symmetry remain unexplained.

Fig. 21. *Make sure that you use tweezers to drop the crystal seeds into the jar – not your fingers*

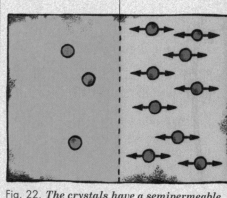

Semipermeable membrane

Fig. 22. *The crystals have a semipermeable membrane which lets some water through in a process called 'osmosis'.*

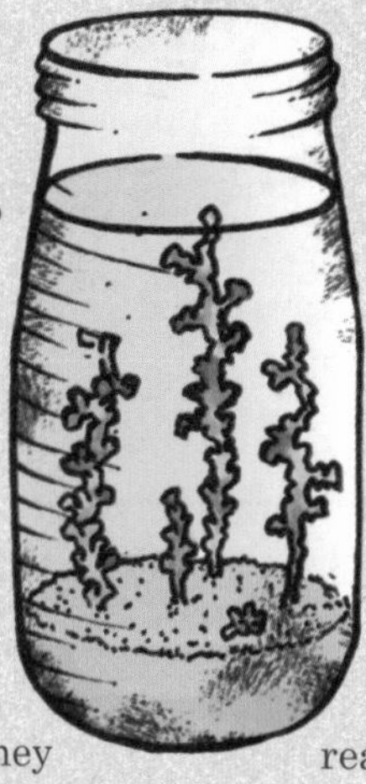

garden. This might be a way to attract younger kids to the project because it'll be easier for them to get going. They're widely available in toy and hobby stores or via the Internet, usually under the name of 'magic rocks'.

How It Works

When you add the chemicals to the water in the first step, they become 'solutes' – basically, substances that dissolve in other substances. By heating the water, you're able to dissolve more of these solutes into the solution than if the liquid was cold. This produces a 'supersaturated' solution. When this is decanted into a saucer and evaporates, it leaves behind the 'seed' crystals. The more slowly you're able to make the supersaturated solution cool, the bigger the crystals you end up with.

Fig. 23. *The crystals will begin to form in your 'garden' almost immediately.*

So how do these 'seeds' become fully-fledged crystal plants? Well, science has the answer. As soon as they hit the sodium silicate solution, the edges of the crystal seeds react to form a precipitate – it's actually a metal silicate gel. This has a semipermeable skin through which some water can pass, courtesy of osmosis. Pressure builds up as this happens, making the skin expand and then burst. When it bursts, it causes a further reaction, creating more precipitate, and the whole cycle continues until all of the 'seeds' have completely reacted.

19 MAKE A LAVA LAMP

Here's a style icon from the 1970s: the lava lamp. They've undergone something of a revival in recent years, but it's much cheaper to make your own.

What You Need

A large red marker pen, a bottle of mineral oil, some green food colouring, two bottles of isopropyl alcohol (one 91 per cent and the other 70 per cent), a pair of rubber gloves, a plastic funnel, a glass bottle with a cap, a tin opener, an empty coffee can, an old-fashioned tin opener with a 'V' end, something to cut tin, a 40-watt light bulb and an old ceramic light-bulb base with mains connector. The size of the can, bottle and base are related, which is why no dimensions are given here. Basically, the can needs to fit over the base and the bottle needs to sit on top of the can.

What To Do

Take the top off the bottle of mineral oil, put on the rubber gloves and unscrew the inky felt tube from the marker pen. Drop it into the bottle and leave it until the oil has changed colour. Take your food colouring and add half a dozen drops to both bottles of isopropyl alcohol. Make a series of 'V' openings round the bottom of the can with the opener, then using your tin cutters, make a light-bulb-sized hole in the middle of the bottom of the tin. Pop the bulb into the fitting and then put the can upside down on top

Make a series of 'V'-shaped holes with a tin opener

of that. Put the funnel into the empty glass bottle and pour in some of the 90 per cent isopropyl, then add some of the mineral oil. Watch it sink to the bottom. Pour in some of the 70 per cent isopropyl and give the bottle a few gentle swirls. Screw the cap onto the bottle tightly and turn on the light.

How It Works

The isopropyl and the oil won't mix – indeed their natural inclination is to separate into layers. However, as the bulb heats the contents of the bottle, the warm oil rises in bubbles. When it hits the top it cools down and sinks. Then the whole thing starts all over again.

Groovy, psychedelic – the ultimate in 1970s living

Fig. 24. *The heat from the light bulb inside the can rises and gets the bottle contents moving to create that authentic-looking lava effect.*

20 THE RESONANT BOTTLE

In this short experiment you'll discover how objects can be made to vibrate in such a way that they actually make exactly the same sound. All you need for this is a couple of bottles of exactly the same shape or size. If you still use glass milk bottles, these will do perfectly.

What To Do

Get someone to hold one of the bottles up against their ear while you stand about 1m (3ft) away. Blow gently across the top of your bottle until you hear it make a tone. As you do, your friend will hear exactly the same tone in their ear – it will just sound further away.

How It Works

It's all due to resonance (where one object is made to vibrate by another), and that's why it's important for both bottles to be exactly the same shape and size. If they aren't then they won't resonate at the same frequency and your friend won't hear the same sound produced in their bottle.

21 TOO MUCH PRESSURE

Everyone knows that you can drink water through a straw, but with a couple of simple adjustments, you can make it impossible, no matter how hard you try. All you need for this is a clean jar with a tightly fitting lid, a hammer and nail, a straw, and some clay to seal it with.

What To Do

First, fill the jar with tap water. Take the lid and use the hammer and nail to make a straw-sized hole in it. Poke the straw through the hole and use a little of the clay to seal round it, then screw the lid onto the jar (the lid must be sealed tightly for the experiment to work). Now try drinking through the straw. You can suck as hard as you like, but as long as the seal is true, you won't be able to drink a drop.

How It Works

Normally, when you suck through a straw the air pressure inside your mouth is less that the air pressure outside. By sealing the jar you prevent the air pressure from acting on the water, so it's impossible to draw water up the straw.

22 WHERE'S HE GONE?

Most of the time, both our eyes look at the same thing and combine the images in front of us into a single 3D picture. But what happens if your eyes are actually looking at different things? Can you trick them into making something that's right in front of you simply disappear? Let's find out.

Fig. 25. *Can you really make your friend's face disappear by rubbing it out with your hand?*

What You Need
A small mirror that you can hold easily in the your hand, a handy white wall (or any large, flat, white surface), and someone else (so you can make them vanish).

What To Do
Sit facing your friend; you should be positioned in such a way that the white wall is on your right and their left. Hold the mirror in your left hand and put the edge against your nose, angled so that the mirror side is facing the wall. Alter the angle of the mirror so that your right eye can only see the white wall reflected in it. Your left eye will still be able to see your friend sitting in front of you. Now, move your right hand up and down the wall on your right and watch what happens. As you move your hand back and forth, it looks as though you're rubbing out your friend's face!

Usually this works a treat, but sometimes one eye is much stronger than the other. If you're having trouble seeing the illusion, simply switch eyes and see what happens.

How It Works
As mentioned above, your eyes normally both look at the same thing. The brain takes these slightly different images and fuses them into a single three-dimensional view of the world. Positioning the mirror so that one eye sees a reflected image of the white wall (or 'nothing') confuses your brain, which is busy trying to put the two images together as usual. Finally, since the brain favours images that move, your right hand appears to be erasing your friend's face, which is static.

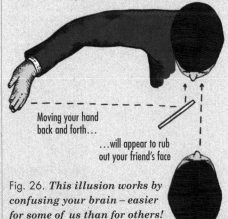

Moving your hand back and forth...

...will appear to rub out your friend's face

Fig. 26. *This illusion works by confusing your brain – easier for some of us than for others!*

23 HOW TO MAKE WATER FLOW UPHILL

Sounds impossible? Guess again. There are various methods for making water flow uphill, such as Archimedes' screw (a simple mechanical device believed to have been invented by Archimedes in the third century BC) and locks (as used on canals). The method used in this experiment is, however, by far the simplest, and unlike these other, man-made techniques, is replicated in nature itself. Read on to find out more...

What You Need

Some plastic wrap (otherwise known as clingfilm), a knitting needle, some sticky tape (variously known as Sellotape or Scotch tape), food colouring, a glass, and some water.

What To Do

Wrap the clingfilm around the knitting needle, then tape the edges and take out the knitting needle to leave a narrow tube made from plastic.

Make sure that the new plastic tube is not bent and is open at both ends, then pour some of the food colouring into a glass that is half-filled with water.

Place your plastic tube into the glass of water. Examine the tube carefully. The dye should climb up the tube until it is above the level of the liquid in the glass. Try the experiment using tubes of varying

Capillary Action

REAL SCIENCE

The scientific principle of capillary action also explains how water travels – against the force of gravity – from the soil upwards through plant stems and into their leaves. Plants contain many vein-like tubes that carry water from the plant's roots upwards to the plant's highest leaves. Larger plants require something called 'transpiration' to keep them cool and move sufficient quantities of water and minerals to where they are required, but let's stop before things start to get too confusing...

thickness (by using thicker and/or thinner knitting needles. You should find that the narrower the tube, the more pronounced the effect will be.

How It Works

Even though we usually think of water as running downhill, it can indeed flow upwards, and this process is called 'capillary action'. This occurs when the adhesive intermolecular forces between the liquid and a solid are stronger than the cohesive intermolecular forces within the liquid. The same effect causes a concave meniscus to form where the liquid is in contact with a vertical surface.

24 THE FEATHER FLAME

When electromagnetic light waves meet an obstruction they spread and bend. This is called 'diffraction', and you can see it in action in this experiment. All you need is a good-sized candle stub on a saucer (or in a candlestick), something to light it with, and a feather.

What To Do
Simply light the candle and let the flame burn for a minute or so or until it's burning steadily. Sit about 1m (3ft) away from the candle and then hold a feather in front of one eye and close the other one. With the open eye, look through the feather and at the candle. No matter how tight the membranes in the feather, you'll still be able to see the candle – or rather you'll see lots of candles spread out in a pattern. You'll also be able to see all the colours of the spectrum.

How It Works
The light from the candle is diffracted by the tiny slits in the feather – the more slits there are in the feather, the more images of the candle flame you'll see.

25 WHY BLACK IS HOTTER THAN WHITE

TAKE EXTRA CARE!

Why do so many athletes wear brightly coloured clothes while people who live in the desert often wear black? For this you'll need an empty tin can, minus the label. You'll also need a lit candle and a pair of pliers.

What To Do
When the candle's burning steadily, hold the can over the flame so that one side on the inner surface of the can becomes 'smoked' and turns  black. Next, find a standing lamp that you can take the shade off, switch it on and hold the can over it by gripping the bottom edge with the pliers. Hold it there for a couple of minutes and then take the can away. Hold it in both hands, making sure one hand is holding the side of the can that's been smoked and the other is holding the 'unsmoked' side.

How It Works
The side that's been coated with black absorbs the heat faster and retains it for much longer, which is why that side of the can will feel hotter. And that desert clothing? It's more to do with the fact that the clothes are loose-fitting, so the wind wicks the heat away more quickly than it can be absorbed.

26 MAKE A MORSE-CODE TRANSMITTER

This dot-dot-dash-dot dot-dash dash-dot-dot-dot electronic gizmo is one of the earliest and most atmospheric of communications devices, conjuring up images of daring, secrecy and danger. It's also relatively simple to build your own.

What You Need

A couple of pieces of cardboard 20 x 10cm (8 x 4in) and another smaller pair of 8 x 2.5cm (3 x 1in), three pieces of wire long enough to stretch between the two transmitters, three shorter pieces of wire about 20cm (8in) long, a 1.5-volt battery, four drawing pins, two LEDs, something to strip the wires (such as a pair of scissors), some tape and a pair of pliers.

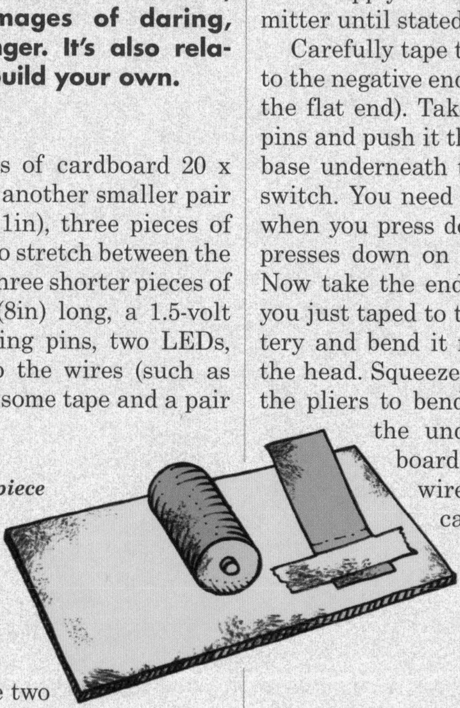

Fig. 27. *Use a small piece of cardboard for the Morse-code 'tapper'*

What To Do

Start by taking the two smaller pieces of cardboard and bending them about 2cm (3/4in) from the end. You can then tape them to the larger pieces of cardboard to act as the switches that actually tap out the Morse code. Tape the 1.5-volt battery to the middle of one of the larger pieces of card with the positive end facing the end of the switch that's taped to the cardboard. See Fig. 27 for the positioning of the switch and the battery. Next, strip about 2cm (3/4in) from the ends of all of your wires so they're ready to be connected.

Most of this experiment concentrates on the piece of cardboard that's got the battery on it, so all the instructions that follow apply to that part of the transmitter until stated otherwise.

Carefully tape two of the short wires to the negative end of the battery (that's the flat end). Take one of the drawing pins and push it through the cardboard base underneath the raised end of the switch. You need to position it so that when you press down on the switch, it presses down on the head of the pin. Now take the end of one of the wires you just taped to the bottom of the battery and bend it round the pin, under the head. Squeeze it tight and then use the pliers to bend the pin flat against the underside of the cardboard base to secure the wire. Tape the LED to the card on the other side of the battery, leaving the connecting wires free.

Fig. 28. *Make sure the drawing pin underneath the tapper lines up with the end of the wire that's attached to it.*

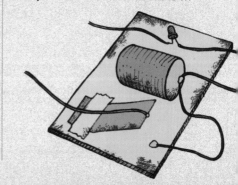

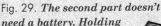

Fig. 29. *The second part doesn't need a battery. Holding down the tapper completes the circuit with the battery on the first station.*

You should have one wire with one end connected to the battery and the other end free. Connect that free end to the negative terminal on the LED (although LEDs look circular, one of the wires comes out of a side that's flat and that's the negative terminal).

You're nearly finished with this part of the transmitter. Push a second drawing pin into the end of the switch so that when you push it down, the heads of the two pins meet. Wrap the end of one of the long wires round the drawing pin and use the pliers to bend it shut. Take one end of the second long wire and tape it to the positive end of the battery. Finally, take one end of the last long wire and connect it to the positive end of the LED. You're now done with the main part of the transmitter.

Now it's time to move to the second piece of cardboard. At this stage you should have already positioned the switch but still have an LED, two drawing pins, a spare short wire and the other ends of the three long wires.

Tape the LED onto the cardboard base. Position the first drawing pin under the switch as you did on the other transmitter, and the second pin on the switch itself so that they'll touch when you push the switch down.

Next, take one end of the remaining short wire and wrap it round the bottom pin. Take the free end of the long wire that's attached to the positive end of the battery and wrap that around the bottom pin as well. Use pliers to bend the point of the pin back to secure the two wires. Attach the free end of the short wire to the positive terminal on the LED, then attach the free end of the long wire that's attached to the switch on the other transmitter and attach it to the negative terminal of the LED. Take the free end of the final wire and hook it round the free pin on the second switch. Bend the pin to secure it. And that's it.

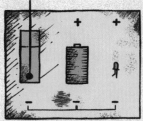

Fig. 30. *The connections for both Morse code stations are simple enough.*

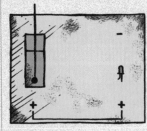

Fig. 31. *You could replace the LEDs with buzzers if you like.*

How It Works

When you press down on the switches, you complete the circuit. This passes the current to the LEDs, which light up. You can substitute these with buzzers if you like, and if you think that the noise would be more fun.

27 THE STARCH TEST

Starch is found in lots of things – not just freshly laundered shirts – but is it possible to find out whether something contains starch using your home laboratory?

What You Need

Try various types of food – for example, a slice of bread, a carrot and an apple. You'll also need a few old saucers or side plates, a medicine dropper, some tincture of iodine, starch powder, a spoon and a glass jar.

What To Do

Start by using the dropper to place five drops of the iodine solution in the bottom of the jar, then add about 5cm (2in) of water and stir the ingredients together. The starch powder is your control – its job is to let you know what starch looks like when you see it. So, put some of the starch powder onto a plate and add a few drops of the iodine solution to it. You'll see the powder turns a very distinctive shade of blue-back, showing that starch is present. Take your other three plates and – giving each sample a wash first – add a slice of bread to one, then a slice of the carrot and then a slice of the apple. Add a few drops of iodine solution to each one and watch what happens.

How It Works

If starch is not present in whatever you're testing, the iodine solution will remain a sort of muddy brown. However, the more starch there is, the stronger will be the change to a rich blue-black colour. On the whole, fruits like apples contain little starch, while cereals contain plenty, as do potatoes and carrots. Washing your samples helps to increase their absorption and so makes the test more effective.

Add iodine solution with a dropper

Fig. 32. Items with more starch in them will change colour and take on a strong blue-black hue like this piece of bread.

28 DECEIVE YOUR EYES

Although most of the time our eyes are reliable interpreters of what's going on around us, they're also easy to fool, as this selection of fun tricks and experiments demonstrates.

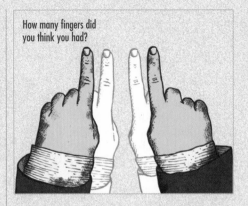

How many fingers did you think you had?

What You Need
A couple of drinking glasses, some sand, a couple of cloth handkerchiefs, a roll of card, and a pair of hands!

What To Do
Start by holding both hands about 20cm (8in) in front of you, with the palms facing away from you. Make them into fists, but leave the index fingers pointing up towards the ceiling. Move your hands together until they're about 5cm (2in) apart and stare in between, rather than at your fingers. After a moment, you'll see extra fingers appearing between them.

As a variation on this, turn your hands inwards so the two fingers are pointing at each other. Leave them about 2.5cm (1in) apart and do the same thing. Your fingers will 'grow' until they appear to meet in the middle.

Next, roll up the card to make a telescope and hold it up to your right eye. Hold your hand up, palm facing away from you, and position it about halfway along the tube. Keep both eyes open but stare hard down the tube and after a moment – you'll see your left hand has a hole in it!

As for the glasses, fill one with sand and leave the other empty. Cover both with a handkerchief and get a friend to pick them both up at the same time. Watch the surprise on your friend's face.

How It Works
Because you're looking between your fingers, into the distance, each eye sees two fingers and this produces four in total. Similarly, with the telescope trick, your eyes mix the two images together (for a more dramatic example of this see page 35). The two-glasses trick is simple. Without really thinking about it, we gauge the weight of an object and prepare our muscles to meet that weight. By making one of the glasses much heavier than the other, we fool our friends' muscles.

Fig. 33. *This experiment will make a hole appear in your hand – and yet it's completely painless.*

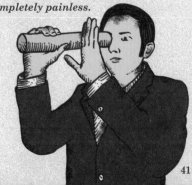

41

29 THE BALLOON HOVERCRAFT

There are a couple of hovercraft experiments in this book. This is the little one, but it's still fun and will teach you a lot about how the big one on page 105 works!

What You Need
A cork, something to make a hole through the cork, some glue, a pencil, a large tin can, a balloon, a small square of hardboard or plywood, a drill and a saw of some description. You might experiment with other materials like Formica and get a better result. Basically, you're after the lightest, stiffest material you can find.

What To Do
Using the base of the tin as your guide, draw a circle on the wood with the pencil and then saw out the shape carefully. Bore a small hole in the centre of the circle about 4mm (5/32in) in diameter, with the drill. Then bore a slightly larger hole through the centre of the cork – say around twice the size of the first one. Be careful when you do this, and think about using a vice if you've got access to one. Glue one end of the cork to the rough underside of the wood (this is actually the top of the 'hovercraft') so it fits over the hole. Give the other side a quick polish with a household wax, or by rubbing a candle on it. Finally, blow up the balloon and stretch it over the top of the cork, put the hovercraft on a smooth surface and watch it go.

How It Works
As the air escapes from the bottom of the balloon and is forced first through the small hole in the cork and then through the smaller hole in the wood, it creates a layer of air between the hovercraft and the ground which reduces friction. This allows the hovercraft to fly across the surface.

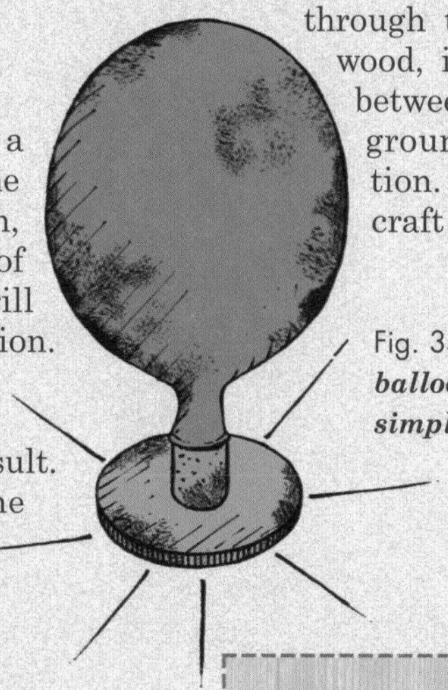

Fig. 34. *Air escaping from the balloon is used to power this simple hovercraft.*

Friction

This experiment is a good example of friction in action. Although the surface of something might look smooth – for example, the polished bottom of our balloon hovercraft, or the surface it's sitting on – in actual fact, it's covered with microscopic imperfections. That means, far from gliding across the floor, it's actually bumping and dragging along. This is friction in action – a force that drags in the opposite direction to movement, whether it's left or right, up or down. It's more useful than it sounds. Without friction we – and everything else – would be sliding all over the place all the time.

③⓪ THE LEVITATING OLIVE

We all know that good things come in small packages, and despite its diminutive stature the olive is an amazing fruit. It makes an excellent lubricant, can be applied to burns and sores as part of a liniment, and is a healthy addition to a balanced diet. More importantly for the shed scientist, you can impress your friends by making one levitate.

What You Need
This experiment requires a minimal amount of equipment – just an olive or two and a large glass (a brandy glass is ideal for this).

What To Do
The challenge here is to get the olive into a glass without touching or picking it up (the olive, that is). Step one is to place the olive on a flat, smooth surface – the top of your shed workbench should be perfect. Next, place your glass over the olive. Slowly slide the glass

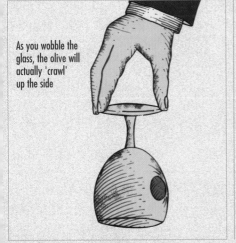

As you wobble the glass, the olive will actually 'crawl' up the side

---

Centrifugal Force
Although a hard concept to explain without using lots of complicated maths, centrifugal force is something that we're all familiar with – it's that sensation of been pushed 'outwards' that you experience on a fairground ride, or in a fast car when it goes around a tight corner. In fact, centrifugal force does not actually exist on its own – it is an equal and opposite reaction (Newton again) to centripetal or 'centre-seeking' force, which is the force needed to move an object in a circle at constant speed.

REAL SCIENCE

---

around the olive, using a swift circular motion. As the olive rolls around, it should start to climb higher and higher up the wall of the glass. Once the olive is spinning around and you're confident that you've got it under control, quickly flip the glass upright – you'll need a good flick of the wrist for this, and it might take a practice run or two to get it right.

How It Works
This trick works because of centrifugal force, which pushes the olive against the side of the glass as it is rotated. The reason that a brandy glass works particularly well is simply that it is thinner around the rim and fatter around the middle, making it more likely that the olive will stay in the glass when you flick it upright.

31 INVISIBLE INK

Ah, an old trick, but always a good one regardless of whether it's part of a wider experiment (and there are several good spy-style experiments to try in this book) or even just to prove that it can be done.

Fig. 35. *Lemon juice makes excellent invisible ink and is a lot more pleasant than raw onion juice!*

What You Need

It is possible to create many kinds of invisible ink using various liquids, including milk, onion juice and baking soda, but this is the least smelly. You'll need a lemon, a knife, something to squeeze the juice into, some paper to write the message on and something to use as a pen (a fountain pen without any ink in it or a toothpick will do). You'll also need a candle and something to light it with.

What To Do

Cut the lemon in half with the knife and squeeze the juice of both halves into a cup or whatever you're using as the receptacle for your 'ink'. Dip the pen or toothpick into the ink and then write your message on the paper. Although it's pretty faint, you should still be able to make out what you've written. Set aside the paper to dry. When it's completely dry, you won't be able to see the message at all – your ink has become invisible. To read the message again, light the candle and then carefully hold the paper over it. After a while the message slowly appears as if written in brown ink. This bit needs practice – you might simply set fire to the paper, but it's easy to judge things correctly after a few goes.

How It Works

It's all to do with the temperature at which different substances start to burn. The lemon juice is basically citric acid, which has a lower burning point than the paper the message is written on. That means it will begin to burn before the paper does. What happens is that the heat makes the citric acid react with oxygen and the acid becomes oxidised, so you can read the message.

Apply a little heat and the message is revealed

32 THE BOTTLE FOUNTAIN

Why is it that we never get tired of watching things that erupt or spurt or otherwise spew stuff forth for no apparent reason? While you're pondering this most universal of questions, have a go at this simple experiment – it's a bottle and it makes a fountain.

What You Need

A glass bottle with a screw top, some food colouring, ink or paint, tap water, a sturdy, straight drinking straw, a needle, some modelling clay, a bowl, some hot water and something to make a hole in the bottle top.

What To Do

Half-fill the bottle with some tap water and add a few drops of colouring. Punch a small, straw-sized hole in the bottle top and then screw it tightly onto the bottle. Poke the straw through the hole and down into the water until it's a finger-width from the bottom of the bottle. Press the modelling clay around the straw where it goes into the bottle top to seal it. The success of the experiment depends on getting a good seal here. Press another small blob of clay onto the top of the straw and then poke a tiny hole in it with the needle. Put the bottle upright in a bowl of warm water and wait. After a short while, water forces its way up the straw and shoots out of the small hole in the clay at the top, like a fountain.

How It Works

It's all to do with expansion. As the warmer water in the bowl heats the air inside the bottle, that air expands, thus taking up more room. That means there's less room for the water, and so it takes the only way out and pushes up and out of the straw. By poking a tiny hole in the modelling clay on top of the straw, you're forcing the coloured water to travel through an even smaller space and that increases the power of the fountain.

Add a little food colouring with a dropper

Fig. 36. *As the air inside the bottle expands, the coloured water is forced out of the top in a fountain-like spray.*

33 EXTRACTING DNA FROM FOOD

This experiment will show you how to extract the very stuff of life itself from food, without any special skills and without using anything more complicated than a kitchen blender. Just how is this possible? Partly because we say you can do it, but more importantly because every single living thing contains DNA and under the right conditions you can actually see it.

What You Need

Some food. Since you're going to be mixing it with water and mashing it up, you can choose pretty much anything – perhaps a vegetable you hate, like purple sprouting broccoli. You'll also need some table salt, cold water, a sieve, a cup, some detergent, meat tenderiser, rubbing alcohol, a couple of test tubes and something long and thin to stir with, like a stick or a straw. Oh, and that laboratory standard – a blender.

DNA

Deoxyribonucleic acid (or DNA) is a stretched polymer of nucleotides (a compound made of sugar, one of the phosphate groups and a heterocyclic base), as everyone knows... just kidding. Every living thing is made up of cells, and inside these cells are the blueprints or instructions that tell them what kind of cells they are and what their job is in the greater scheme of things. The molecule that contains these blueprints is DNA. Close up, it looks like a bendy ladder with a twist in it – this is called a 'double helix' structure – with each rung made of two of the four letters in the so-called DNA alphabet: A, C, G and T. These stand for adenine, cytosine, guanine and thymine. The DNA strand is arranged in different ways to form genes that tell your cells to make the proteins defining the way each particular cell functions. Think of them as instructions that tell your body what to do at the most fundamental, cellular level.

Fig. 37. *Yes, you too can extract the stuff of life from everyday foods using ordinary household items.*

Fig. 38. *Sieve the mix to get rid of the gunk, and use the liquid that's left for the experiment.*

What To Do

Take the purple sprouting broccoli (you need about enough to sit in the palm of your hand) and put it in the blender. Add a healthy pinch of salt and a cup of water. Put the lid on securely, switch to 'high' and blend for 20 seconds. Next, pour the gooey green mixture through the sieve and into a cup. Look to see how much is there. Estimate what one-sixth of the amount of liquid is and add

that amount of detergent to the cup. Swirl it around to mix the two together and then leave the whole lot to stand for 10 minutes.

Decant some of the mixture into a test tube until it's about half-full, and then add a pinch of the meat tenderiser and stir very gently. Now you need to add some rubbing alcohol (the same amount as the broccoli mixture) by pouring it carefully down the side of the test tube so that it settles on top, rather than mixing with what's already there. Just take your time when you pour it. Amazingly, DNA will rise up from the broccoli mixture and into the alcohol in thick strings that you can then hook out with the straw – but don't eat it.

How It Works

By putting the broccoli, water and salt into a blender, you separate the cells from each other. The next step is to break down the membranes around those cells so you can get at the DNA inside. Detergent is very good at this (specifically, it removes the lipids and proteins to 'reveal' the DNA within). The reason it happens so quickly is because of the meat tenderiser, which contains enzymes that are useful for speeding chemical reactions and in this instance are used to slice the DNA away from the various proteins in the mix that would otherwise still be 'protecting' it. Finally, by adding alcohol you're providing the remaining DNA with an environment it prefers to the watery green soup that's left at the bottom of the test tube, so it rises to the top and gathers together in strings.

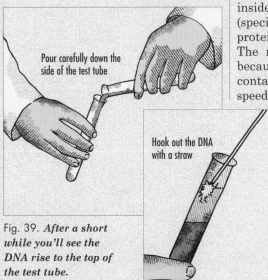

Pour carefully down the side of the test tube

Hook out the DNA with a straw

Fig. 39. *After a short while you'll see the DNA rise to the top of the test tube.*

34 HOW DOES YOUR GARDEN GROW?

You've probably seen plants growing on the television, when they have been filmed using time-lapse photography. Here's how you can measure how fast your garden grows for a lot less effort, and a lot less money.

What You Need
A plant pot and a little compost, a fast-growing seedling (most kinds of beans are good), a sheet of stiff card, a couple of hardback books, a drinking straw, some cotton and a large drawing pin.

What To Do
Plant the seedling according to the instructions on the packet, cover with compost and give it a good drink of water. As soon as the first shoot appears, you're ready to start measuring. Position the card upright between the two books and then push them together to hold the card in place. Next, tie one end of the cotton carefully to the shoot and tie the other end to the bottom of the straw. Pin the straw vertically to the card as shown in the illustration here. As the shoot grows, it will pull the bottom of the straw upwards. By returning at the same time each day and making a mark on the piece of card, you can measure how quickly your plant is growing. Try experimenting with different types of plant, and position them in different parts of the house – both in and out of direct sunlight – to see how much the environment affects their growth. Although everyone knows that plants need sunlight, actually seeing the difference it makes to plant growth at first hand may surprise you.

How It Works
This simple measuring device relies on the plant itself for power, but once you've set it up properly, you need only return once a day to mark the measurement on the card. Expect plants that have plenty of sun to grow more quickly than those that don't. This is because plants are drawn to the light so they can photosynthesise sunlight and grow more quickly.

Fig. 40. *As the plant grows, it will pull the straw round and you can use the movement to measure the growth.*

Tie thread round plant when it first sprouts

35 THE EGG STRIPTEASE

This project may sound a bit strange, but if you follow the instructions properly, it will reveal something about an everyday object – an egg – that you've never seen before.

What You Need
The ingredients are simple. You need a few ordinary eggs, some white vinegar, a watertight container big enough to hold the eggs, a cover for the container, a plate and a large, egg-friendly spoon.

What To Do
Gently lay the eggs in the container, making sure that they're not touching each other. Pour the vinegar into the container until the eggs are covered (but don't actually pour it over the eggs themselves). Put the cover on the container and wait a few minutes. Look inside and you'll see that bubbles have begun to form on the outside of the shells – this is the first clue that something out of the ordinary is happening. Stick the container in the fridge for a day or so.

Once they've been left for long enough, take the container out of the fridge and – as carefully as you possibly can – remove the eggs with the spoon and put them very gently on the plate. This needs a steady hand because, amazingly, the shells have started to disappear, which means the only thing holding the egg together is the gelatinous outer membrane. Be prepared to lose one or two at this stage! Tip the vinegar down the sink. Put the eggs back into the box and cover with fresh vinegar, put the lid back on and put them back in the fridge for another 24 hours.

Now you can take the eggs out and examine them. No shells, very squidgy, very bizarre.

How It Works
The shell of an egg is made of calcium carbonate, while the vinegar contains acetic acid. The acid breaks the shell into calcium, which drifts away, and carbonate, which turns into bubbles (carbon dioxide), leaving the wobbly outsides.

In Vinegar Veritas

Vinegar ('vin aigre', meaning 'sour wine') has been around in one form or another for about 10,000 years, but it's only recently that it's extraordinary properties have begun to be recognised. It's acid that enables vinegar to do so many things, from providing powerful flavours to preserving pickles and cleaning household surfaces. White distilled vinegar is good for drains, and makes windows sparkle. It also softens hard paintbrushes, can keep your hands soft if you've been working with garden lime, will make blankets fluffy if added to a wash, and soothes stings, sore throats and burns.

REAL SCIENCE

36 COLOURFUL FIRE

There are a number of companies that will happily sell you specially treated wood, pinecones or other flammable items that are designed to turn your home fire into a colourful light show. Don't waste your cash on these – burn much less money on this experiment instead. As always, remember that you're dealing with chemicals and that they should be treated with care.

What You Need

This depends entirely on the colours you're after. Remember, for example, that sodium produces a mainly yellow flame that's pretty much indistinguishable from what you get when you burn normal stuff, so you want to go easy on that. Start with calcium chloride or potassium chloride, which will produce orange and purple flames respectively. Pinecones are a good thing to use as the fuel because they have lots of nooks and crannies to absorb the chemicals. You'll also need a bucket and some water.

What To Do

Pour enough water into the bucket to give the cones a thorough bath, then stir in some of your chosen chemical until it's completely dissolved. Aim for a ratio of 4 pints of water to 1/4lb of chemical (and mix them together outside, while wearing rubber gloves). Add the cones and swish them about so they become well coated, then leave them to soak overnight. Remove them carefully and set to dry. Add them to your fire sparingly at first until you can see how effective they are!

How It Works

Essentially you're producing a simple firework without having to bother with the most difficult bits – the propellant and binder. If you want to combine different colours, remember to prepare them separately and then add them to the fire from the different batches. You can buy most of these chemicals at your local hardware store, or even large supermarkets if you look in the laundry or cleaning sections. Make sure that you wear gloves and goggles when you're mixing these chemicals in case any of them splash.

Fig. 41. *Pinecones are great for absorbing the solution because they've got such a lovely, knobbly surface.*

37 CREATE A CLOUD

Of course, most of us would rather see the back of the clouds, but the way they are created is interesting in itself, especially when they're the DIY kind. You'll need a large glass jar, some water, a metal tray of some description (a typical baking tray will do the job), some ice cubes, a room where you can block out the light, and a torch.

What To Do

Pour some warm water into the jar (about 2.5cm or 1in is enough) and then pop the metal tray on top of the jar. Pour enough ice cubes into the tray to cover the bottom, and leave it for a few moments. Carry the lot carefully into a darkened room and place it on a flat surface. Take the torch and shine it through the glass, into the jar. You'll see that you've managed to make a cloud.

How It Works

When the warm air inside the jar meets the cold bottom of the tray, it condenses to form a cloud. You'll also see drops of water collecting on the underside of the tray, like raindrops.

38 RUST AND WATER RISING

Although we spend much of the time worrying about rust getting to our bikes, cars or garden tools, the speed at which rust occurs is actually rather interesting, as this experiment demonstrates.

What To Do

The first thing to do is to get a good palmful of steel wool and jam it firmly into the bottom of a glass. This is the only tricky bit to this experiment, because you must make sure that the wool's going to stay in place, even when you turn the glass upside down. When you're happy, add a few drops of vinegar to speed up the rust reaction and then place the glass upside down in a bowl of water. As the steel wool in the glass rusts, the water rises up the glass.

How It Works

Rust is actually made up of iron, water and oxygen. As the steel wool absorbs oxygen from the air, the pressure inside the glass decreases. As it tries to equalise the pressure again, the air outside pushes the water from the bowl into and then up the glass.

Condensation inside the jar causes a cloud to form

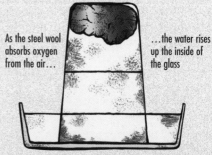

As the steel wool absorbs oxygen from the air…

…the water rises up the inside of the glass

39 MAKE YOUR OWN RADIO

This is a classic project that everyone can enjoy, even if you've never delved into the world of amateur radio before. Who would have thought you could put together a radio that works from a few inexpensive components and then run it forever without batteries?

What You Need

The list of ingredients for your radio is a combination of components you can buy at any decent electronics store, along with stuff that you have lying around the house. For a start, you'll need a strong plastic bottle about 7.5cm (3in) across and about 15cm (6in) long. These measurements are approximate; as long as the bottle is strong enough to support all the wire you're going to wrap around it, it will work fine. Next, you'll need about 15m (50ft) of enamelled copper wire between 18 and 22SWG (standard wire gauge) – they'll know what this is in the shop, even if you don't. Then, get something called a germanium or 1N34A diode. This may be slightly harder to track down, but your local store may be able to order it or recommend an alternative if you tell them what it's for.

Obviously you need something to listen to the radio with, but this doesn't have to be anything more complicated than an old-fashioned single earpiece (the technical term is a piezoelectric earphone). Don't try and use modern stereo headphones, because they probably won't work. Then get a set of insulated alligator leads and 15m (50ft) or so of stranded insulated wire – again, your electronics shop will be able to advise

you here. Finally, you'll need something to make holes in the plastic bottle, a pencil, some electrical tape and some sandpaper. (Note: we've tried to steer clear of soldering in this book, but if you know what you're doing it's the best way of attaching some of the components described in this and other experiments.)

What To Do

Start by punching four small holes in the side of the bottle, two near the base and two near the top, as shown in fig. 42. Ideally, the holes should be about 13mm ($1/2$in) apart. Take one end of the enamelled wire and thread it through the two holes nearest the bottle top. Pull it through about 25cm (10in) and then bend it back and run it

Fig. 42. *You create the coil for the radio by wrapping the wire around the bottle.*

Fig. 43. *When the wire has been wrapped around the bottle and pencil, stage one of your crystal wireless set is complete, and you're ready to remove the pencil.*

through the two holes again – this double loop will help to keep the wire in place.

Now start to wrap the wire around the bottle, starting from the end where the top is (and where you've just tied the loop) and working towards the bottom of the bottle. When you've done half a dozen turns, get a pencil, lay it parallel to the bottle and loop the wire once round that. Then make another six turns around the main bottle before looping round the pencil again. Keep going until you've reached the other two holes you punched at the bottom of the bottle. By now you should have a bottle wrapped in wire with a pencil held in place alongside it by smaller loops that occur every six turns of the main wire.

To finish off, thread the end of the wire through the two holes, and then loop it back through them again as you did with the first two holes near the top of the bottle. Cut the end of the wire so that there's about

15cm (6in) sticking out. You've now completed stage one of your radio.

The next step is to slide the pencil out. Now you should see the bottle wrapped in wire with a line of smaller empty loops down one side. In a short while each one of these will represent a radio 'station'. What you need to do now is remove the coating from these small loops, using the sandpaper. You don't need to touch the wire on the main bottle at all, just get rid of the insulation on the smaller loops.

Sandpaper the ends of the longer bits of wire so that you'll get a good connection when you come to connect them to the rest of the radio. You need to connect the loose end of the wire nearest the bottom of the bottle to the diode by twisting them together and then securing everything with a bit of electrical tape.

Next, you need to attach the earpiece, so cut off the connector from the end and uncoil the twisted pair of wires. Strip the insulation off the ends. Attach one end to the other side of the diode and the other end to the loose

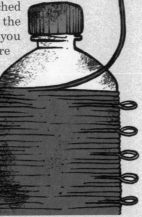

Fig. 44. *The small loops made by the pencil will become your radio 'stations'.*

wire coming off the large coil at the bottle-top end.

Take one of your alligator leads and clip it onto the end of the wire coming off the top of the bottle (where you've already attached one wire from the earphone). Attach the other end of the alligator lead to something like a cold water pipe to make sure your radio is earthed properly. Take the second alligator lead and clip one end onto one of the small loops you created along the side of the radio and clip the other end onto your wire.

If you listen through the earpiece now, you'll hear a radio station broadcasting through your radio. If the signal sounds weak, you can improve it by making sure your aerial wire is as high as possible and that the cable is stretched out, rather than lying looped on the floor. You can also try listening to other stations by moving the second alligator clip to a different small loop on the side of the radio.

How It Works

Most radios are powered devices, designed to make a copy of the sound that's transmitted by a radio station so that it can be amplified and played back. On the other hand, a crystal radio (so-called because early models used a galena crystal) is able to use the power of the radio signal itself, receiving the electromagnetic energy and then passing it on to the crystal detector. This samples the radio wave and then turns into a faint audio signal that you can hear in the earpiece. Tuning a crystal radio set is an inexact science, but that's all part of the fun.

Having said that, there's nothing more frustrating than spending time building a wireless that doesn't seem to receive anything. If you're having problems, you probably need to increase the length of your aerial wire.

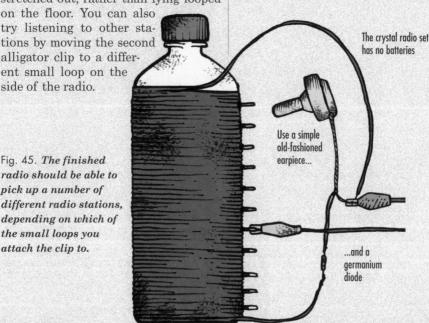

The crystal radio set has no batteries

Use a simple old-fashioned earpiece...

...and a germanium diode

Fig. 45. *The finished radio should be able to pick up a number of different radio stations, depending on which of the small loops you attach the clip to.*

40 THE EXPLODING BAG

Our old friend carbon dioxide provides the power for this particular experiment, which is actually extremely noisy! You should probably try this either outside or over an empty sink – when the bag goes bang, it makes quite a mess.

What You Need
One of those plastic sandwich bags with a seal that you can press shut (a bit like a zip). You'll also need some water, an ordinary cup, a tablespoon of baking powder, some vinegar, a pair of scissors and a paper towel.

Fig. 46. *Once you've given the bag a shake, make sure you drop it somewhere sensible, like an empty sink – it's going to make a mess when it explodes.*

What To Do
Test the bag for leaks by pouring in some water, sealing it and giving the bag a good shake. If all is well, empty the bag. Next, take a paper towel and cut it into a 13cm (5in) square. Pop one and a half tablespoons of baking soda into the middle and then fold it up as shown in our illustration. Next, measure out half a cup of vinegar and a quarter of a cup of warm water and pour them into the bag. Seal the bag until there's just enough room to slip the folded paper towel (and its baking-soda payload) inside. Pop the paper towel in and quickly finish sealing the bag. Give it a little shake and then drop it into an empty sink. The bag will inflate and explode with a loud bang.

How It Works
When the acidic vinegar reacts with the baking soda it produces carbon dioxide, which quickly expands. Because the bag is sealed this rapidly expanding gas can't escape; the pressure builds until it's so powerful that it bursts the bag. If the bag doesn't burst, try adjusting the levels of the three ingredients. If you want to give yourself more time before the bomb explodes, add a couple of extra folds to the paper towel – this will stop the elements combining quite so quickly.

41 CONFUSE YOUR NOSE

There are so many wonderful flavours in the world that it's hard to believe that in reality the taste buds on your tongue can only distinguish four types: sweet, sour, salty and bitter. All the rest – as this experiment shows – is actually down to your nose.

Hold the slice of pear under your nose while you eat an apple – what does it taste of?

What You Need

Start with something simple – an apple and a pear. If you want to go on and make the pie, you'll need the following: a pre-made pie crust (when you see what's inside the pie, you're not going to want to waste your time making any pastry), two cups of water, $1^1/2$ cups of sugar, $1^1/2$ teaspoons of cream of tartar, 20 unsalted crackers, a little cinnamon, one teaspoon of butter, an ovenproof pie dish, a saucepan, a wooden spoon and a measuring cup.

What To Do

Cut the pear in half and hold it under your nose, then eat the apple. Because the pear smells more strongly than the apple and because they have similar textures, it feels as though you're eating a pear. If you want to fool your taste buds even more, make the pie. Follow the instructions to prepare the pie crust, and pre-heat the oven to about 200°C. Put two cups of water in the saucepan and bring them to the boil. While you're waiting, mix the sugar and cream of tartar together and then add them to the boiling water a bit at a time, stirring as you go. Next, break the crackers into large, rough pieces and stir them in. Boil for a couple of minutes without stirring. Pour the mixture into the pie crust, add a little cinnamon and smooth the butter evenly over the top. Bake in the oven for 15 minutes. Tell someone that you'd like them to try a slice of your famous apple pie and see what they say.

How It Works

The cream of tartar in the pie is acidic, and when it combines with the sugar, water, pastry, crackers and cinnamon, most people will believe that they're eating a genuine apple pie!

Fig. 47. *A pie that tastes of apples, yet doesn't have any apples in it – what's going on?*

42 MAKE YOUR OWN WEED KILLER

If you're struggling to find a weed killer powerful enough to clean up your garden yet gentle enough to leave your plants to grow healthily, look no further.

What To Do

All you need is a few basic household items to make these powerful and effective weed killers. Try a tablespoon of gin, another of cider vinegar, a teaspoon of liquid soap and a quart of hot water. Mix them together and spray directly onto the weeds. Alternatively, a gallon of distilled vinegar mixed with a tablespoon of liquid soap and a large cup of salt will also do the trick.

How It Works

Weeds don't like the high acidic content of ingredients like vinegar and lemon juice, while the detergent that's used in these recipes helps 'stick' the acid to the weeds. Remember that you should only use vinegar products that are specifically labelled as herbicides.

43 BOTTLE MUSIC

From the haunting tones of South American panpipes to the altogether more prosaic grunts of an old-time jug band, blowing across the top of a hollow tube is an effective way of making music, as this experiment shows.

What To Do

All you need for this is a series of glass bottles of the same size and shape – milk bottles will work great. If there's a piano or other instrument nearby that you know is in tune, so much the better. Use as many or as few bottles as you like. If you've got access to twelve, you can play a complete octave, but the principle will still work even if you've only got three or four bottles. Line them up and then fill them with different amounts of water. If you blow across the top of the bottle you'll hear a note, and if you're able you can adjust the amounts of water in each bottle until they're roughly in tune with the piano.

How It Works

The more water there is in a bottle, the less space there is and the faster the air vibrates, making a higher note. Those bottles with less water in them will make deeper notes.

44 GLOW-IN-THE-DARK GHERKINS

You've probably never considered the possibility of combining the power of electricity with the pungency of a pickled vegetable in order to see what happened. Nevertheless, that's what this project is all about, and you'll discover that vegetables too have a part to play in bringing light to the dark corners of the world.

What You Need

As with all experiments to do with electricity, you need a healthy respect for something that – while being a supremely useful tool – also has the capacity to hurt you very badly. That's why you should read the Safety First section on pages 14–16 before you do anything, and then come back here.

OK, this experiment is best conducted in a shed, garage or workshop where there's power. Then you'll need a couple of ring stands and two clamps to fit on them (you'll need to track down a supplier of chem lab equipment for this) two screws, a strong pair of scissors (or

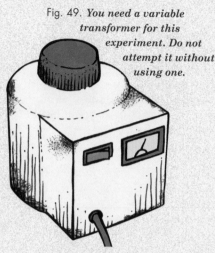

Fig. 49. *You need a variable transformer for this experiment. Do not attempt it without using one.*

wire cutters), some electrical tape, an ordinary mains extension lead and a variable transformer. Oh, and various kinds of pickled vegetables, especially gherkins, dill pickles and pickled onions. If you can't get hold of a variable transformer, leave this experiment alone until you can; it's important that you're able to control the current precisely.

What To Do

Start by attaching the clamps to the ring stands and then setting them up so that they're roughly 30cm (1ft) apart. You may have to move them closer together, depending on the size of the vegetable you're using. Next, cut the end off the extension cord – not the end with the plug on it. Then separate the wires, strip off the insulation at the end of the brown (live) and blue (neutral) wires and wrap each one round the head of one of the screws. Keep it in place with the tape. For safety, tape up the end

Make sure that the ring clamp is secure...

...and that the clamp is on a flat surface

Fig. 48. *You'll need a pair of ring clamps to hold the gherkins in position.*

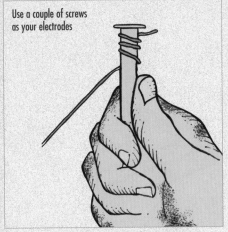

Use a couple of screws as your electrodes

Fig. 50. *Strip the insulation off the wire and carefully wrap it round the top of each screw.*

of the unused wire (the green/yellow striped one). Next, secure each screw in a clamp so that they're facing each other. Then, gently press the vegetable onto one of the screws and then onto the other. It's very important that the two screws don't meet in the middle or they can produce an arc – similar to that used in welding – which is highly dangerous.

With the variable transformer set to '0', switched off and unplugged, you can now plug the extension cord into the transformer, and then plug that into a wall socket. Switch on the transformer and get a friend to turn off any lights if you have them. Slowly turn the voltage up to 120. Your vegetable will begin to exhibit some interesting new characteristics – it will start to spit, fluid will drip out, it'll smoke and then, finally, give off a definite glow. Remember to switch off and unplug everything before trying to replace the pickle with a new one.

How It Works

In fact, it's probably not the pickle that's doing the glowing at all, rather it's the gas that's given off when the current heats the liquid inside the pickle. The screw electrodes can easily heat the pickle to boiling point (as evidenced by the discharge and the smoke), but it's not until the gas begins to cool that you see any light. How does it cool while the current is still switched on? The water vapour produced by boiling coats the screws and prevents conduction, which lowers the temperature before allowing heat to build up again.

There's some evidence to suggest that those pickles with a higher sodium – salt to you and me – content will burn for longer and more brightly, but as with all good experiments, you need to try these things out for yourself.

Fig. 51. *Stand well back when you do this experiment, and be prepared for much dripping, fizzing, smoking and fizzling to go along with the illuminations.*

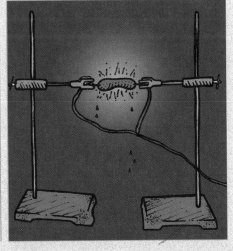

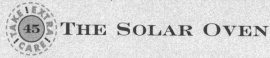

45 THE SOLAR OVEN

Cooking outdoors sounds like a wonderful idea, but is usually accompanied by unpleasant but necessary wood gathering and fire lighting. Not so with the solar oven, which uses the heat of the sun to cook your food to perfection.

Fig. 52. *A family-sized pizza box is just the job for this experiment.*

What You Need

A large, flattish empty box – a family-sized deep-pan pizza box is just the job. You'll also need foil, a piece of clear plastic sheeting, a sheet of black card, duct tape, some scissors, a pen, some glue and something like a stick or a heavy-duty straw to hold the lid of your oven open.

What To Do

Mark a square on the top of the pizza-box lid. This should be about 2.5cm (1in) in from the sides. Use the scissors to cut along the front and sides of the edges you've marked out, but leave the back attached so it can act as a hinge. You should then be able to bend it back like a lid. Cut a square of foil to the size of the lid and glue it in place, smoothing out any wrinkles as you go.

Next, open the proper lid of the pizza box and cut a square of plastic slightly larger than the hole you cut in the box lid. Tape it tightly to the underside of the proper lid. Cover the inside of the box with more foil and then cut and fit a square of the black card into the bottom of the box. Put your food into the oven, close the proper lid and then prop the top lid (the one you cut) so that it reflects the maximum amount of heat into the oven.

How It Works

The foil reflects the heat of the sun onto your food, or pot if you're using one. The key element, however, is the black card, which absorbs the heat and then – thanks to conduction – transfers it to the cooking food. In bright sunlight, you can get up to 90°C using this technique!

Fig. 53. *Beware: your oven can get extremely hot and must be treated with caution.*

46 THE MAGIC MOVING TUMBLER

This little trick will baffle onlookers, as you move a glass tumbler down a slight gradient without ever touching it. The scientific principles behind it are simple, but applied to the real world they can be used to move things that are much, much heavier. Read on for more...

What You Need

For this experiment, you need just a few simple items: a plastic-covered tray or an old glass picture frame, something to prop up the tray/picture frame at a slight angle, a glass tumbler, a candle, and a small amount of water.

What Do To

Take the plastic-covered tray or old picture frame and tilt it very gently. Then prop at this angle. Wet the rim of the glass tumbler and place it upside down on the glass or plastic surface. (The slope must be gentle enough for the tumbler to stay in position.) Then light the candle and gently heat one side of the tumbler. As soon as the side of the tumbler becomes slightly warm, it will start to move by itself – as if by magic!

*Fig. 54. **The experiment allows you to move a glass tumbler as if by magic!***

Hover History

A hovercraft rides on a cushion of air in much the same way as the glass tumbler in this experiment. But the principle of the floating tumbler was first used for a new kind of railway shown at the Paris Exhibition of 1878. The train had runners instead of wheels; they floated on water-filled rails to give a smooth ride. No manufacturer has yet taken up this idea, although versions using magnet levitation are becoming increasingly popular. The principle is much the same, but the results are even more effective because there is no contact between the rail and the carriage, and therefore no friction to slow it down. Public versions are already operating in Germany and China, and in the case of the latter, the trains can reach an amazing top speed of 430km/h (267mph).

How It Works

The secret of how this trick works is as follows. Heat from the candle makes the temperature of the air inside the tumbler rise. Hot air rises, so the tumbler is lifted up from the surface. The water on the rim acts as a seal so that the air inside cannot escape. Because a thin cushion of water supports the tumbler, its own weight sends it 'floating' down the slope.

47 MAKE A PERISCOPE

A periscope isn't just useful in a sub-marine. You can use it to spy on your friends and it can even allow you to see over other peoples' heads if they're blocking your view of a concert or sporting event. And as you'll see – it's simple to build.

What You Need

A piece of stiff card about 30 x 20cm (12 x 8in). It needs to be really stiff, too – your best bet is to try a proper modelling shop and get the stuff that's designed to be scored and bent into shape, rather than relying on something like the back of a stiff-based envelope. You'll also need a couple of mirrors about 6cm (2¹/₂in) square, some sticky tape, a pencil, ruler, and a sharp modelling or utility knife.

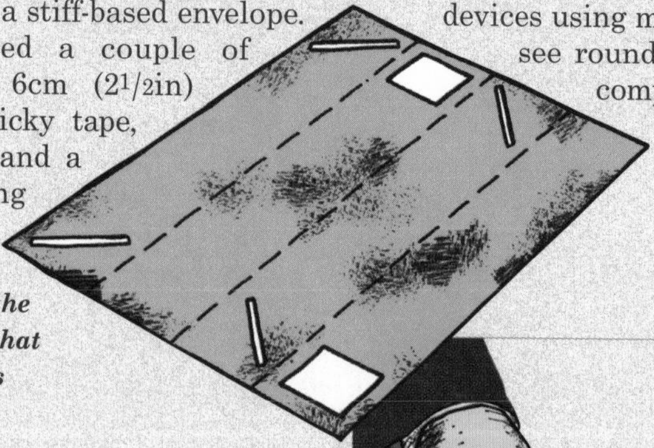

Fig. 55. *Mark out the card carefully so that the periscope folds together neatly.*

What To Do

Divide the card lengthways into four equal columns using the pencil and ruler, and then score down them with the knife so that you can bend the card-board into a square tube. Next, cut a couple of square view holes into the cardboard, slightly narrower than the width of one column, as shown in Fig. 55. Then, using the utility knife and ruler, cut four slits at a 45-degree angle; these should be wide enough to take the two mirrors. Bend the cardboard to

form a square tube and then tape it securely. Slide the two mirrors into posi-tion, making sure you get them facing the right way. And that's it. The periscope will only work well in natural light, but it'll give you a unique per-spective on the world.

How It Works

When the light from whatever you're looking at hits the top mirror, it's reflected down the periscope tube to the bottom mirror so you can see it. There's no reason you couldn't use the same principle to build more sophisticated devices using multiple mirrors to see round corners or more complex obstacles.

Fig. 56. *Slide the mirrors into the diagonal slots – making sure they're the correct way around.*

48 THE STATIC ELECTRICITY FLEA CIRCUS

This experiment uses the power of static electricity to produce a fun effect that makes it look as though grains of rice or bits of cereal are jumping up and down like fleas in a circus.

What You Need
Nothing very much: a sheet of clear plastic about 30cm (12in) square and 3mm (¹/8in) thick. You'll also need a piece of woollen cloth, a large piece of white paper, some grains of uncooked rice or puffed-rice cereal, and four half-sized tins (these must all be the same height).

What To Do
Put the paper on a flat surface and place one can at each corner. Rub the plastic sheet vigorously with the woollen cloth for about a minute. Scatter the 'fleas' on the paper and then place the plastic sheet on top of the cans. Your 'flea circus' will spring (or rather jump) into action.

How It Works
The rice and the plastic have an equal number of positive and negative charges, making them electrically neutral. By rubbing the plastic with the cloth, you give it a strong negative charge. When the plastic gets near the fleas it polarises them, pulling positive charges to the top and pushing negative ones to the bottom. Then the negatively charged plastic attracts the positively charged tops of your 'fleas', so they appear to jump towards the plastic. If they touch it, they'll get a good kick of negative electricity from the plastic, turning the top of the flea neutral, but – remember the flea was neutral in the first place – actually giving it a slight excess of negative charge, which pushes the flea away from the plastic. This continues until the negative charge drains off, at which point the electrically neutral flea is ready to start all over again.

Fig. 57. *When you rub the plastic sheet with wool it creates a strong negative charge which will get the 'fleas' jumping.*

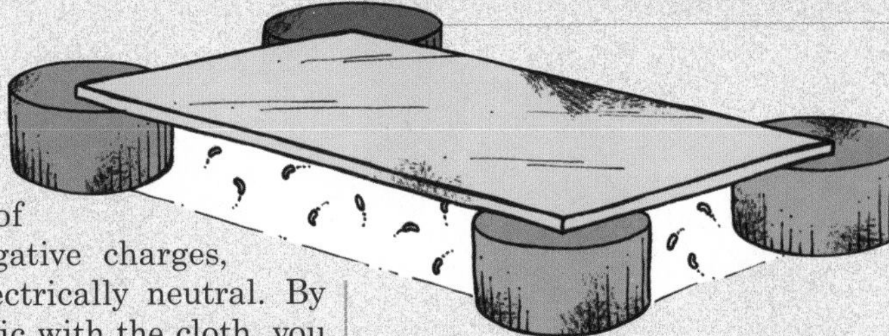

Fig. 58. *The cycle of charging and discharging static electricity should enable your 'fleas' to put on a good show.*

49 TURN A RAINBOW WHITE

How easy is it to make someone's eyes play tricks on them? Try this simple experiment and you'll find how you can make all the colours of the rainbow disappear.

What To Do
Use a compass to mark out a circle on some card, and then cut it out. Use a ruler to divide the circle into seven segments, and mark these out in pencil. Paint the segments the following colours in order: red, orange, yellow, green, blue, indigo and violet. Put the card to one side and let it dry. When it's ready, make a small hole in the middle with the scissor point and push the pencil through. Using the pencil, spin the disk as fast as you can. Look at the colours and watch.

How It Works
Basically, your eyes can't keep up with the colours as they spin round. To make things easier, they merge into a single colour – white – and that's what your brain thinks it's seeing.

50 CANDY CHROMATOGRAPHY

Hidden inside a simple sweet you'll discover the secrets of chromatography and capillary action. For this experiment you'll need some blotting paper, a tube of sweets (flying-saucer-shaped ones are good), some water and a large plate.

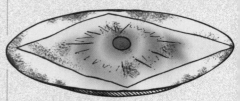

What To Do
Cut the blotting paper into a plate-sized shape and put it on the plate. Put a sweet in the middle of the paper. Dip your finger in the water and let it drop onto the sweet. Keep dripping water until it's good and wet and there's a damp circle of about 5cm (2in) across around the sweet on the blotting paper. Let the blotting paper do its stuff, and in a while you'll be able to see rings of different colours around the sweet.

How It Works
The sugar coating around the sweet contains colouring, which dissolves when you add water. The blotting paper uses capillary action to draw the water (and the colouring) out through the paper. Because the colouring is made of different inks, these are drawn out at different speeds by the blotting paper. The result is very basic chromatography – a way of separating and analysing complex mixtures.

51 A FILM-CANISTER ROCKET

This experiment is very simple, but great fun. It's based around an ordinary canister of the type used to hold camera film. White and transparent plastic film canisters usually work better than the black canisters with the grey tops, because the caps fit on more tightly. Follow the instructions below and you'll be able to launch a rocket that goes 3m (10ft) in the air!

What You Need
A plastic film canister (the type of container that most 35mm camera film comes in), the lid that comes with the film canister, an antacid tablet (such as Alka-Seltzer), and some water.

What To Do
The canister will go quite high, so it's best to perform this experiment outside.

Remove the lid from the film canister and put one antacid tablet inside; then add a teaspoon of water.

Fig. 59. *Add the antacid tablet, followed by a teaspoon of water. It is important to replace the cap quickly once the water is added.*

It's important to do the next two steps quickly: first, put the cap on and make sure that is on tightly. Second, put the canister on the ground with the CAP SIDE DOWN and STEP BACK at least 2m (6ft).

About 10 seconds later, you will hear a pop and the film canister will launch into the air!

Tip
If it does not launch, wait at least 30 seconds before examining the canister. Usually the cap is not on tight enough.

The canister should reach a height of 3m (10ft)

How It Works
When you add the water it starts to dissolve the Alka-Seltzer tablet. This creates carbon dioxide and pressure begins to build up inside the film canister. As more gas is made, more pressure builds up until the cap it blasted down and the canister is blasted up. This system of thrust is how a real rocket works, and you can improve the rocket by adding fins and a nose cone made out of paper.

Fig. 60. *Once the mixture is ready, place the canister cap-side down on the ground, stand well back, and watch it fly into the air.*

52 HOW STEADY IS YOUR HAND?

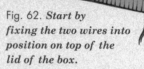

This uses a simple circuit to create an amusing game that all the family will enjoy playing. Once you've understood the basic principle, you can make it as simple or as complex as you like so that everyone can join in the fun.

What You Need

An ordinary shoebox for starters, a little bulb holder, a 3.5-volt bulb (you can get these from any electrical store), a 4.5-volt battery, three short lengths of electrical wire, pliers, two lengths of thicker wire (a couple of old metal coat hangers will probably do the trick), a pair of scissors and some electrical tape. The exact length of the wire is in direct proportion to the size of the box you use – take a look at the final illustration and you'll see what this means.

Fig. 61. *Once again, ordinary household objects wait to be transformed into something that's fun and unusual.*

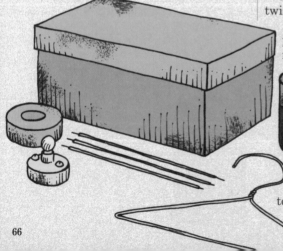

Fig. 62. *Start by fixing the two wires into position on top of the lid of the box.*

What To Do

Use your scissors to make two small wire-sized holes in the top of the shoebox at either end, and then make a light-bulb-sized hole in the middle of the box lid as shown in the illustration. Make a fourth small hole in line with the bulb hole and at right angles to one of the wire holes. Take one of your pieces of thick wire and twist the end round to form a loop and then turn it round the main wire several times so you've got a ring at the end.

Next, take the other thick wire and twist it into a series of bends, a bit like a rollercoaster. These can be as simple or complex as you like, but remember that they have to be far enough apart to allow the wire ring you've just made to pass round them – you want to make it difficult, but not impossible. Poke one end through one of the holes you just made in the lid, bend it back, and tape it to the underside securely. Take the other end and poke it through the wire ring and then through the second hole in the box lid. Wrap the end of one of the short lengths of electrical flex around it, tape them to secure the connection, and then tape both to the lid of the box.

Next, put the bulb in the holder and attach the other two pieces of electrical wire to it using the built-in screw terminals. Thread one of the electrical wires through the remaining hole in the lid and twist it round the handle of the loop wire. Don't tape it. Attach the remaining flex from the bulb holder to one battery terminal. You should then have one wire end still free – attach that to the other battery terminal. Test the device by touching the loop to the main twisted wire. The light bulb should come on. If it doesn't, check your connections and try again. When you've got it working, secure the light-bulb holder to the underside of the lid with tape and then put the lid on the box.

How It Works

As long as the loop of wire doesn't touch the main 'rollercoaster' wire, the bulb won't light up. However, as soon as you touch the two wires together, you complete the electrical circuit and the bulb lights up. That's why you don't tape the last electrical wire to the loop of thick wire – this allows you to undo it when you're not using the toy, so as not to waste the battery. For more fun, you can replace the little light bulb with an equally small buzzer, which is also available from most electrical stores.

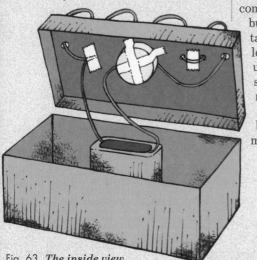

Fig. 63. *The inside view of the box, showing the connections.*

Fig. 64. *So, how steady is your hand? When you accidentally touch the main wire, the light will come on and you've lost.*

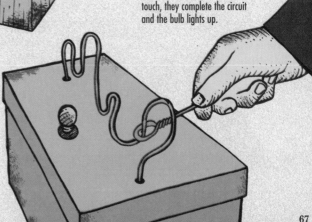

When the two pieces of metal touch, they complete the circuit and the bulb lights up.

53 MAKE A VOLCANO

No, it's not an actual volcano – that wouldn't be a good idea at all. Instead, build this special-effect volcano, which shares many of the characteristics of the real thing without any of the attendant dangers.

What You Need

Baking soda, flour, salt, cooking oil, warm water, food colouring, an empty plastic drinks bottle, some dishwashing detergent and some vinegar. This is going to make a mess, so you'll also need a large baking tray to place the finished volcano in.

What To Do

There are two stages to this experiment. The first is the cosmetic, mountain bit, and how long you spend on it is up to you. If you mix together six cups of flour, two cups of salt, four tablespoons of cooking oil and two cups of water, you'll get a good, flexible dough that you can mould around the empty bottle and into the shape of a mountain. Experiment with the amounts if you want a bigger or smaller version. You

Real Eruptions

In real life, volcanic eruptions vary depending on the make-up of the magma and how much water is present. Super-hot temperatures and lots of water will result in faster-flowing magma, while the amount of dissolved gas is in direct relation to the force of the eruption itself. So it's not too dissimilar from our table-top version. By the way, the Smithsonian in the US lists about 1500 active volcanoes worldwide (or 1501 with yours).

can then paint it, or add little fake trees, or do what you like to make it more mountain-like. When you've finished, put it on the baking tray.

Next comes the chemistry. Get a jug of warm water and add some red food colouring, then pour the mixture into the bottle until it's about seven-eighths full. Add about six drops of the detergent, then two tablespoons of baking soda. Pour the vinegar slowly into the top – and be ready to jump back when your volcano becomes active.

How It Works

The vinegar reacts with the baking soda to produce carbon dioxide. This builds up pressure in the bottle and since there's only one way out, the liquid inside – now full of bubbles thanks to the carbon dioxide and the detergent – shoots out of the top. The red food colour simply makes it look like lava.

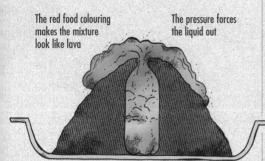

The red food colouring makes the mixture look like lava

The pressure forces the liquid out

Make sure you've got a good-sized bowl to collect all that 'lava'

54 MAKE FIREWOOD FROM PAPER

It's perfectly possible to take old newspaper and turn it into something that will burn rather like wood, at the same intensity and producing the same amount of heat. All you need is a pair of scissors, some newspaper, a large, shallow container, and some foil.

What To Do
First, wait for a sunny day. Cover the inside of your dish with the foil and then cut the newspaper into container-sized pieces. Soak these in warm water 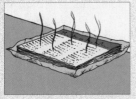 to get them nice and soggy, and then pack them into the container. As you go, give each layer a squeeze to get rid of any excess water and pour it out of the container. Keep going until you've got a couple of centimetres (about 3/4in) of wet newspaper in there. Lift the foil lining out with the paper still inside and leave it out in the sunlight until it's dried.

How It Works
Wetting the paper makes it stick together – it's almost like reconstituting the original wood – and tiny holes between the layers let oxygen permeate through to help keep the fire going. If you're the impatient kind, stick the container in the oven for an hour on a medium-high heat – though this probably uses more energy than you're saving in the first place!

55 THE MAGIC BALLOON

This is one of the oldest tricks around, but young kids still find it entertaining and it's a great way to show off the power of static electricity. All you need for this experiment are some balloons and a few volunteers wearing woollen jumpers.

What To Do
Start off by simply getting them to rub the balloons against their jumpers for about 30 seconds. When they've done that, get some of them to just let go of their balloons and watch as they float up to the ceiling and stick there. Get the others to hold their balloons against the wall and let go. In both cases, the balloons will 'magically' stay where they are.

How It Works
When you rub a balloon against the jumper, it removes negatively charged electrons. When you let go of the balloon or hold it against the wall, the uncharged ceiling or wall attracts the charged balloon and keeps it there until the charges equal themselves out. This can take hours.

56 MAKE A HOT-AIR BALLOON

Yes, you could simply blow up a balloon, tie it off and let it float off into the sky, but it's hardly the same thing, is it? In this experiment, you'll produce a model of a real hot-air balloon, burner and all!

What You Need

Ten sheets of tissue paper of about 25 x 40cm (10 x 16in) – you can adjust this depending on how large you want your balloon to be. Then you'll need a 2B pencil, some scissors, paper clips, glue, a 70cm x 6mm (28 x 1/4in) length of art card or poster board, an empty coffee tin, some thin wire, a pair of tin snips, a file, a large ball of cotton wool, some rubbing alcohol, a lighter or some matches, and a taper.

What To Do

Start by creating the shape of the balloon using the sheets tissue paper – note the words 'tissue paper' and remember that it tears easily. Take your paper and stack one sheet on top of the other, then fold it lengthwise. Use a few small paperclips along the fold to keep it in position and then, using the illustration as a guide, draw the wavy line from top to bottom. Then take your scissors and cut along that line. Carefully remove the paperclips and open out the tissue paper. You'll now discover that you've got eight sheets of paper that are shaped roughly like a bishop's hat. You can throw the other bits away. What you want to do is glue all these pieces of

paper together like the bellows of a concertina. Take two pieces and glue their right edges together, then take another piece and glue the left edge of that to the left edge of the top piece of paper. Take another piece and glue its right edge to the right edge of the top piece of paper, and so on. When you've glued the last piece of paper, open the balloon out and glue the final two edges together. You can either scissor off any excess paper at the top you don't need, or tie it tight with some cotton thread. Although it's good to seal the balloon as much as you can, it doesn't have to be completely airtight. Now leave it to dry.

Bend the card into a circle, gently open the bottom of the balloon and adjust the card – as if you were measuring someone's neck – so that it will fit the space. Hold it between your finger and thumb, and allow an overlap that you can glue. Remove the card

Fig. 65. *The metal cross at the bottom will open the 'mouth' of the balloon.*

and glue it, cutting off any excess. When it's dry, glue the outside edge and fix it in place round the bottom of the opening in the balloon.

Measure across the hole in the bottom of the balloon and then cut two wires to length so that you can position them to form an 'X' shape – you'll need to make holes in the cardboard collar, poke the wires through and leave enough at each end to fold it down, thus locking it in place. Now take your big wad of cotton wool and wrap some of the wire round it, leaving enough to form a little hook at the top (you'll hang this off the 'X' you just made). You're going to use the coffee can as your balloon launcher, but in order to get in and light the cotton safely, you need to cut a square out of the side of the can as shown in the illustration. And that's it.

Wet – but don't soak – the cotton with the alcohol, and then hang it from the 'X' using the wire. Take it and the can outside. Light the taper with the match and then light the cotton. After a short while the balloon will rise majestically into the air.

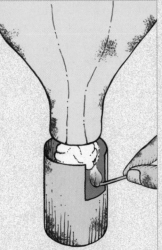

Fig. 66. *Be very careful when you light the ball of cotton wool.*

How It Works

Hot air rises. As the air inside the balloon is heated by the burning cotton wool it inflates the tissue paper. Because the air inside the balloon is enclosed, it remains hotter than the air outside the balloon and so allows the balloon to rise through the air. It will only drift back to earth when the cotton wool goes out and the air inside cools again, allowing the weight of the balloon itself to take over.

Remember: this thing is burning, so choose your launch site with care and perform the experiment responsibly.

You can have lots of fun with this experiment. Try running a balloon competition with friends. Everyone can paint their balloons in their favourite colours (or the colours of their favourite team), and you can then see whose balloon goes highest and furthest. If you get kids involved, they'll need supervising because there's fire and inflammable stuff involved.

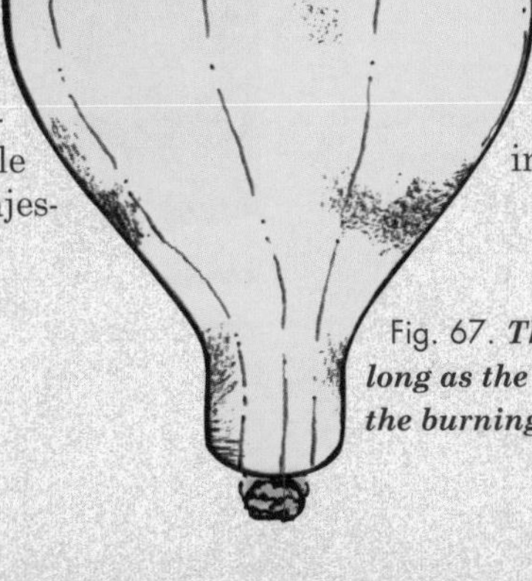

Fig. 67. *The balloon will fly for as long as the air inside it is heated by the burning cotton wool.*

57 BAFFLE YOUR BRAIN

Naturally, this experiment will be easier to do with some people than others, but you'll find that it has an almost 100 per cent success rate. In fact, if you're the betting kind, you could almost make some money with this because it's such a sure-fire winner.

Fig. 68. *How can something that seems so simple be so difficult? Try it for yourself and see if you can do any better.*

You Will Need
You can do this trick by yourself, but it's more fun with a friend because you get to enjoy their frustration as they keep trying – and failing – to do it properly. In either case, all you need is an ordinary table and chair, a piece of paper and something to write with. That's it.

What To Do
This is basically a variation on the old 'patting your head while rubbing your stomach' trick, except this one is much better and much more likely to work. Tell your friend to sit in the chair and then raise one foot off the ground. Get them to start making a circular motion with their foot. When they've got the movement going at an even regular speed, ask them to sign their name on the paper – it's important that they try to scribble their usual signature, rather than just writing out their name. They'll find this almost completely impossible; either their foot will stop moving entirely, or it will start to follow the shape of their signature. Alternatively, their signature will begin to follow the rotation of their foot and they'll draw circles on the paper. It's fantastically funny to watch.

How It Works
Although humans are rather good at multi-tasking and doing several complex activities at once, this is only possible when each action uses a different part of the brain. When you ask your brain to do two entirely different things which happen to use the same part of the brain to co-ordinate them, it simply can't cope.

58 SEE YOURSELF IN AN EGG

As any shed scientist knows, eggs are fabulous things – useful both in the kitchen and in the laboratory. Here, we'll show you how you can treat an egg so that it turns into a mirror.

What You Need
A candle and something to light it with. One of those little tea lights in its own foil cup is good for this. You'll also need an ordinary uncooked chicken egg (though shed scientists know there's no such thing, really), some kind of tongs to hold the egg over the flame (barbecue tongs work well), and a big glass about half-full of tap water.

What To Do
Pop the candle on a flat surface and light it. When the flame is established, take the uncooked egg and pick it up carefully with the tongs. Hold it over the flame. After a short while you'll notice it starts to change colour – it gets darker. Move the egg around with the tongs until all or at least most of the shell has changed colour, and then let the egg cool. It probably needs at least five minutes, so give it ten to be on the safe side. Once it is cool, gently lower the egg into the glass of water and watch what happens. The surface of the eggshell takes on the characteristics of a mirror.

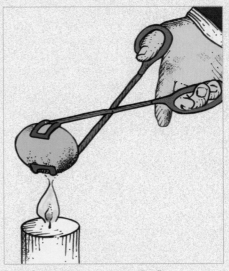

Fig. 69. *Hold the egg over the flame with a pair of tongs – try to ensure it blackens evenly.*

How It Works
It's to do with the soot. Soot is carbon dust, which forms when something burns but doesn't actually combust. The carbon in the soot pushes the water away and preserves a layer of air between it and the eggshell. This results in the unexpected reflective effect, which will persist until the air is completely dissolved into the water – something that will take a couple of minutes to happen.

Fig. 70. *When it's placed in the glass, you should be able to see your face reflected in the silvery surface of the egg – at least until the effect wears off.*

59 FRANKLIN'S BELLS

Depending on how you look at it, this was either one of the first electric motors or a brilliant – if rather dangerous – way to predict when a lightning storm was coming. This version works in the same way, and is fun and easy to build. It's named after Ben Franklin, by the way, printer, scientist and politician.

What You Need

Not any lightning, thank goodness. Real lightning is much too dangerous to mess around with, so for our source of electricity we're going to use static from a television set. You also need a couple of tin cans, some kind of stick (a pencil or lollipop stick will do), a paperclip, cotton thread, some scissors, aluminium foil, some sticky tape and a couple of thin wires about 1m (3ft) long. Last but by no means least, it's a good idea to wear some rubber gloves while conducting this experiment.

What To Do

Make sure the TV is switched off, and then put the cans on top of it. Pull off some thread (you want just less than the height of a can) and tie one end around the paperclip and the other around the middle of the pencil or lollipop stick. Then place the stick across the two tin cans like a little bridge, so that the paperclip is suspended between them. It shouldn't touch the top of the TV; if it does, wind some of the thread round the stick until it doesn't. You should also try to position the stick/thread/paperclip so that the clip is exactly in the middle of the space between the two cans.

Strip any insulation off the wire ends with the scissors and then tape the end of the first wire to one of the tins. Take the other end and tape it to something metal, like the leg of a table or a chair, and tape one end of the other wire to the second tin. Tear off a large square of foil. This should be large enough to cover the screen of your TV, unless you've got one of those over-the-top home-cinema jobs. Carefully tape the free end of the second wire to the front (i.e. the side that's facing away from the TV screen) of the foil in one corner. Ready? Switch the TV on and wait. After a while, the suspended paperclip will begin to swing between the two tin cans.

Fig. 71. *The original Franklin's Bells was designed to detect lightning storms – yours will detect static electricity.*

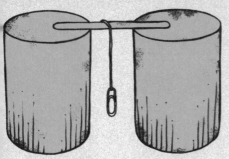

Fig. 72. *A paperclip suspended between two cans will act as the clapper for your set of 'bells'.*

How It Works

In the original experiment, Franklin was serious about detecting lightning, so rather than a TV, one of his wires was connected to a lightning rod that was fixed on the top of his house, while the other wire went to an iron water pump – not something that you should try at home. Some of the electrons used by your TV to create the picture are absorbed by the foil, which sends them to the first can, giving it a neutral charge. The paperclip (which starts out as neutral) picks up this charge and, because like repulses like, gets pushed away and towards the other can. Since the other can is positively charged, when the paperclip hits it picks up a

positive charge and is repelled again back towards the other tin, where it picks up a negative charge, and so on. This process continues until there's no more static electricity left to move the paperclip. When that happens you can turn the TV off. Make sure you're not touching either of the cans as you do, because you'll get a little shock. In the original experiment Franklin used bells rather than tin cans, which were designed to ring when lightning approached – hence the name of the experiment.

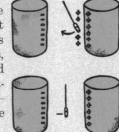

Fig. 73. *The paperclip will swing between the two tin cans until there's no more static electricity left to discharge.*

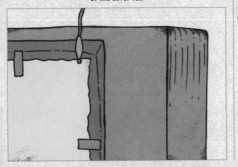

Attach a clip to the edge of the silver foil

Fig. 74. *Make sure you attach the end of the other wire to something like a metal chair leg.*

60 HOW WINGS WORK

Common sense tells you that just as heavy boats ought to sink like giant stones, heavy aeroplanes shouldn't be able to get off the ground, let alone fly. One of the reasons that they can is down to the ingenious way designers utilise air pressure to keep them up.

What You Need

A sheet of ordinary photocopier paper, a pencil, a drinking straw, a couple of straight sticks (disposable chopsticks are good for this) a pair of scissors, a ruler, some sticky tape (or glue if you prefer) and some cotton thread.

What To Do

Measure off a 20 x 10cm (8 x 4in) piece of paper and cut it out with the scissors. Fold it over so that the short edges touch, but don't crease the folded end. Stick the two short edges together. At the other end, make a fold in the paper. Don't make it exactly in the middle; instead, judge where to fold the paper so that it's flat underneath but with a gentle curve on the top as shown in the illustration. Next, use the scissors to make a small hole in both the top and bottom bits of paper so you can pass a straw through them as shown. It's important that the hole is small enough so that straw doesn't move up or down – if it does, glue it in place. Run a piece of thread through the straw and tie the sticks to each end, top and bottom. Hold the sticks out in front of you as far apart as the cotton allows and then swing the whole lot from side to side. As you do, you'll see that the wing rises up the cotton thread.

How It Works

The secret is in the wing shape. By having wings that are flat underneath and curved over the top, designers make sure that the air has further to travel over the top of the wing, thus lowering the pressure underneath and helping to keep your plane in the air.

Fig. 75. *The shape of an aeroplane's wings is crucial to actually keeping it up in the air, as this experiment demonstrates.*

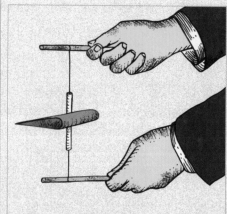

61 A SIMPLE STEAMBOAT

This cute little experiment demonstrates one of the oldest and most effective forms of locomotion – steam – and will provide you with a small but perfectly formed steamboat.

What You Need

A half-sized plastic washing up liquid bottle, one that you can cut easily with scissors. You'll also need a tea light (one of those squat little candles that comes in its own metal cup) and some soft copper tubing of the type plumbers use. You should be able to get it from a hardware store. The important thing is that it should be 3mm (1/8in) wide rather than the more common 6mm (1/4in), which is too heavy for the boat. You should try to get the store to cut it to length for you. The exact length depends on how big your boat is, but you should be able to gauge it roughly by looking at the illustration.

What To Do

Cut the bottle in half lengthways, having first removed the top. Then take the copper tubing and bend it into the shape shown in the picture – it helps to have something small and circular to bend it around. Make two holes in the stern (these should be smaller than the circumference of the tubing) and then push the two ends through. The soft plastic will grip the tubes and help keep the water out. Note that the tubing needs to bend down from the candle, up again as it goes through the back of the

boat, and then down again as it goes under the water. Fill the tubing with water by putting one end in the water and sucking on the other end, and then position the candle as shown under the double bend in the pipe. Light the candle and wait. After a short while the boat will set off.

How It Works

The heat from the candle boils the water in the tube, producing steam, which shoves the water out and pushes the boat forward. The steam condenses to form water when it hits the empty (colder) parts of tubing, creating a vacuum and sucking the water back into the tubes. In this way, the boat can keep going for as long as the candle burns.

Fig. 76. *Make sure that the copper tubing you get is small and light enough.*

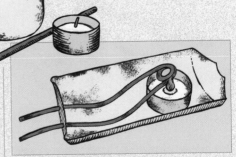

Fig. 77. *Steam provides an extremely effective propellant for a small, light craft such as this one.*

62 PREDICT THE WEATHER

Apparently, you can use seaweed to tell whether or not it's going to rain. Quite how this is supposed to work is a bit of a mystery – but it's one that you can explain courtesy of this easy and fun experiment.

What You Need
A hair from your head, a lollipop stick (any excuse to eat a lollipop), some tape, a drawing pin, a square of stiff card, and a couple of heavy hardback books.

What To Do
The first thing to do is wait for a dry, sunny day. Got one? Good. Now, eventually you're going to stand the card up vertically between the two books, so you need to imagine that there's going to be a top and bottom to the card. Take the hair and tape one end to the top. Take the other end, pull it tight and tape it to the lollipop stick – about 13mm (1/2in) from the end. Pin the other end to the card using the drawing pin so that the lollipop stick and the hair are at right

angles to each other. Remember that the hair needs to be stretched tight and that the pin needs to be in firmly enough to hold the stick in place, but loose enough so it can rotate. The stick is going to be your weather barometer. Draw a picture of the sun next to the end of the lollipop furthest away from the drawing pin. Then, watch what happens to the pointer over a period of a week or so. As the weather gets wetter, the barometer will rise. (You can draw a miserable cloud at its highest point if you like).

How It Works
It's all to do with the amount of moisture that's in the air. When conditions are humid, then the chances of rain are increased. And as the moisture piles up in the air, the hair absorbs it and becomes limp. Just like seaweed, in fact.

Fig. 78. *Using a human hair to predict weather is just as good as using seaweed – and not so smelly.*

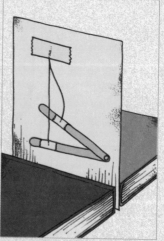

Make sure the hair is taut when you tape it to the card

Fig. 79. *The lollipop stick needs to be able to rotate freely round the drawing pin for this to work.*

63 HOME-BREW INSECTICIDES

As all gardeners know, the only things that love their plants more than they do are the insects that enjoying eating them. Unfortunately, the cure – insecticide – is sometimes so unpleasant that you don't want to use it. Try these natural remedies instead.

What To Do

To get rid of black spot from roses, use a tablespoon of baking soda, a teaspoon of washing-up liquid and a gallon of water. Aphids hate a mix of vegetable oil (one cup), a tablespoon of washing-up liquid and a cup of water. Moles hate jalapeño peppers even more. Beer in a shallow

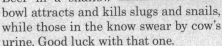

bowl attracts and kills slugs and snails, while those in the know swear by cow's urine. Good luck with that one.

How It Works

Like most insecticides, natural versions interfere with the insects' nervous systems, disrupting them until they can't tell the difference between real stimuli and imagined ones; if they eat plants that have been sprayed the reaction occurs more quickly. Remember too the relative size of insects – it doesn't take much to upset them. Beer and slugs? They don't get drunk and pass out – they drown because they can't crawl out of the beer trap.

64 HOW TO MOVE METAL THROUGH ICE

Did you know that it's possible to move a piece of metal right through an ice cube and out the other end without breaking it in half? All that's required is a bottle with a cork in it, an ice cube, a short piece of metal wire and a couple of large nuts and bolts.

What To Do

Wait for a cold day, then take everything outside and find a nice sheltered spot out of any wind. Tie the nuts and bolts to each end of the wire so you can balance them on the end of your finger. Next, take the ice cube and put it on top of the cork. Carefully balance the wire-and-bolts contraption on top of the ice cube and watch what happens. The wire will move down through the ice without breaking it, eventually emerging at the bottom where it will rest on the cork. The ice cube will remain intact.

How It Works

Pressure. The weight of the bolts causes the wire to press down, melting the ice as it goes. As it sinks, the ice above it re-freezes because it's no longer being subjected to any pressure.

79

65 LIGHT A TAPER WITHOUT USING A MATCH

This experiment is all about the way plants are able to turn sunlight into oxygen via the process known as photosynthesis. It's a kind of double experiment, which starts by producing the oxygen and then demonstrates its presence in a simple but dramatic way.

Fig. 80. *Pond plants are perfect for this experiment because they don't need any soil in order to start photosynthesising.*

Add the water to the main container

What You Need
A friend to help with lighting and holding the taper, a good-sized glass jar, a heavy glass funnel, a test tube (this needs to be wider than the spout of the funnel), some water weed (if you can't get this from a pond or a friendly neighbour, try your local garden centre, which will probably carry a selection of inexpensive pond plants). You'll also need some matches and a taper. This experiment works better – or at least more quickly – on a sunny day.

What To Do
To start with, you need to get these plants making oxygen, so start by dropping them into the bottom of the jar and then place the funnel on top of them. It doesn't matter if the plants are completely enclosed by the funnel; in fact, we need to have a few of them sticking out so that when we add water it circulates properly. (This is one of the advantages of using pond plants, by the way – they prosper in water and don't need to root anywhere in order to photosynthesise.)

So, pour water into the jar until it's about 2.5cm (1in) below the top, then fill

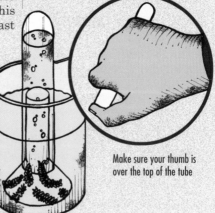

Make sure your thumb is over the top of the tube

Fig. 81. *Within a few hours, the pond plants will do their work and begin to produce oxygen, which gradually pushes the water out of the inverted test tube.*

the test tube with water, too, so that there's no air inside it. Then, in one smooth motion (this is the tricky bit) you need to position the test tube over the top of the funnel spout as shown in the illustration. Take the whole lot and put it in the sunshine and then leave it there. Come back in an hour and you'll notice that things are afoot, specifically that gas bubbles are making their way from the plants up the funnel, into the test tube and onto the surface. Make a mental note of the water level in the test tube and leave it 'cooking' for a few more hours. When you come back you'll notice that the water level has dropped. Let it carry on until the test tube is completely empty of water.

Remove the test tube and keep your thumb over the end. Get your friend to light the taper and let it burn for a moment. When it's caught properly, blow it out and then remove your thumb from the end of the test tube and poke the taper into it – the taper will re-light immediately.

How It Works

Plants use a substance called chlorophyll as their 'engine' for growth. Chlorophyll is able to convert sunlight into food for the plant by photosynthesis. Crucially for this experiment, photosynthesis involves taking in carbon dioxide and producing oxygen. So, as the pond plants absorb light from the sun, they produce enough oxygen to force the water down the test tube and into the main body of the jar, leaving behind a 'tube' of oxygen. Fires are simple things and need only heat, fuel and oxygen to burn. The first two are already present in the taper, so by adding the third, the flame re-ignites. Were you to set the experiment up in exactly the same way but leave the jar in a dark cupboard, nothing would happen. Without sunlight, there's no photosynthesis, which means that the plants don't create any oxygen.

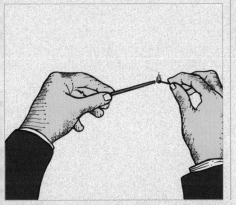

Fig. 82. *As always when you're using matches or any kind of naked flame, take care and treat the taper with respect.*

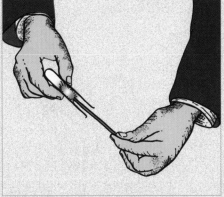

Fig. 83. *Once you've blown the taper out, pop it straight into the test tube and it will re-light, seemingly all by itself.*

66 MAKE AN ELECTROMAGNET

We've looked at the power of magnetism elsewhere in this book, but an electromagnet is a rather different breed of push-me-pull-you. In these three experiments you'll discover how to create an electromagnet and then see it in action.

What You Need

Start with 60cm–1m (2–3ft) of insulated wire – tell your local hardware store what you're planning to do and they'll point you in the right direction. You then need a larger battery than usual – something like 6 volts will work well – because otherwise it's going to drain so

Fig. 84. *An electromagnet differs from a conventional magnet in that it only works when an electrical current is available.*

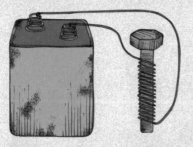

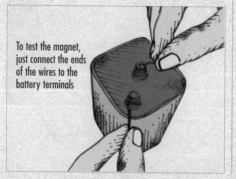

To test the magnet, just connect the ends of the wires to the battery terminals

quickly that you'll be forever replacing it. After that, all you need to complete the first experiment is a large metal bolt and some paperclips. If you'd like to go on to do parts two and three, you'll need a conventional bar magnet and a cardboard tube. The inside of a roll of toilet paper is good, but the inside of a roll of kitchen towel is even better because it's longer and will take more twists of the wire. Finally, you'll need an ordinary small washer.

What To Do

Strip the insulation off the ends of the wire – you don't need to take very much off, just enough so that you can attach it to the terminals on the battery. Then, leaving 15cm (6in) or so free at the end, start by looping the wire around the bolt in a series of tight turns. You want to get as many loops as possible without them actually overlapping, so that when you've finished, you can't see the bolt at all, just the coils of wire. When you're done, hook up the wire ends to the terminals on the battery. Then move the paper clips near to the wire-covered bolt and watch what happens. The clips are attracted to the bolt – congratulations, you've created an electromagnet. Take one of the wires off the battery and the clips won't be attracted any more. (By the way, don't be concerned if the electromagnet gets warm – just disconnect it and let the thing cool down before carrying on.)

Want to try the second part of the experiment? Reconnect the wire to get the electromagnet working again, then take your conventional bar magnet and move the north side nearer to one end of

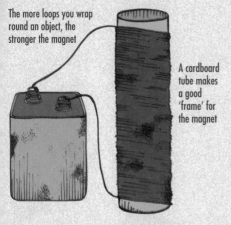

The more loops you wrap round an object, the stronger the magnet

A cardboard tube makes a good 'frame' for the magnet

How It Works

Conventional magnets are always magnetic – they're 'on' all the time, if you like. That's why they're sometimes called permanent magnets. Electromagnets, on the other hand, only work when electricity is present, because a moving electrical charge generates a magnetic field around itself. That's why the loops are important – the more you have, the stronger your electromagnet will be. Being able to switch the magnetic properties of something on and off like this makes electromagnets fantastically useful, and they're used in all sorts of devices from car starter motors to doorbells.

the bolt. If the two are attracted to each other, you've found the south end of the electromagnet; if they repel each other, it's the north end.

Finally, try this. Disconnect the wires and unwrap the bolt. Take the cardboard tube and wrap the wire round it exactly as you did the bolt, making sure that you get as many turns of wire as you can round the tube. Then take the small washer and put it part of the way inside the cardboard tube. Connect one end of the wire to the battery and quickly touch the other end to the other terminal and you'll see the washer disappear inside the tube! For fun, you can also try placing the washer on a flat surface and then holding the tube/coil contraption over it in one hand and touching the second wire momentarily to the second connector with the other. See if your electromagnet is strong enough to pick the washer up.

As far as using the conventional bar magnet is concerned, opposite poles attract each other, which is why the south end of the electromagnet is attracted to the north end of the conventional magnet.

When you wrap the wires around the cardboard tube you need more turns to generate the same amount of magnetic 'strength' than you would if you wrapped it around a metal object like a bolt. That's because the atoms in the bolt line up to form what's called a 'domain', which strengthens and amplifies the electromagnetic field.

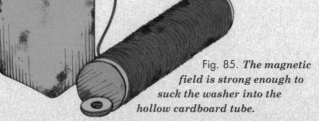

Fig. 85. *The magnetic field is strong enough to suck the washer into the hollow cardboard tube.*

67 SOLAR POWER

You'll find a couple of other experiments elsewhere in this book that can also lay claim to being about solar power – the difference being that in this one, you get to set things on fire. Don't worry, it's all perfectly safe and a graphic demonstration of the power of the sun.

What You Need
First and foremost, you need a sunny day to make this work. After that, all you want is a plain glass bottle, a length of cotton thread, a small, heavy object to use as a weight (something like a bolt would do the trick) and a cork of the right size to fit in the top of your bottle. Oh – and here's the giveaway – you'll also need a magnifying glass. If you like, you can use this experiment to play a trick on a friend – in which case you'll also need one of those.

What To Do
Tie one end of the thread around your weight (using a bolt makes this easy). Then tie the other end round a cork, measuring the length of the thread so that when the cork is in the bottle, the weight hangs suspended 4 or 5cm (about 2in) from the bottom of the bottle. Take it outside and ask your friend to make the weight fall to the bottom of the bottle without touching it. When they can't – since it's impossible – take out the magnifying glass and position it so that the sun's rays are concentrated on the thread. It may take a few minutes, but eventually the thread will smoke and then catch fire. When it does, the weight will fall to the bottom of the bottle.

How It Works
The rays of the sun are easily hot enough to start a fire if they're focussed properly using a magnifying glass. Obviously you'll have to adjust the position of the glass so as to get the smallest possible point of light burning the thread, as this will concentrate the heat better. Never mess about with a magnifying glass in the sun – you can get a nasty burn.

Fig. 86. *Be careful with the magnifying glass – if the sun can generate enough heat to burn the thread, it can generate enough heat to burn you.*

68 MEASURING EXPANSION

We know that heat makes things expand, but it can be difficult to gauge exactly what's going on. To measure expansion all you need is a short length of wire, some tongs, a heat source capable of making the wire red-hot, and a piece of paper.

What To Do

Hold the wire over the heat source using the tongs – obviously, do this very carefully. When the wire glows red hot remove it from the flame and lay it on the piece of paper. The wire will burn through the paper, leaving a mark for its entire length. Once you can see the mark on the paper, gently poke the wire with the tongs so it rolls away from the mark. Leave it to cool. You'll be able to see that there's a clear difference between the length of the mark made when the metal was hot and its length when it's cool again.

How It works

Metal expands when it's heated and returns to its original dimensions when it cools. As the metal absorbs heat, its atoms get a boost of energy and move around more freely, shunting other atoms around and taking up more space.

Carefully heat the wire until it glows red-hot...

...then place it on the paper

69 THE FLYING MIRROR

Strictly speaking, it's not the mirror that does the flying but your good self, thanks to this simple trick that uses reflection to mess with the way you perceive what's in front of you.

What To Do

All you need is a big mirror, a low, sturdy table, a friend to hold the mirror safely in place and another friend to observe the trick. When you've completed it successfully, you can take turns until you've all had a go. All you need to do is lean the mirror against the low table with the reflecting side facing away. Once it's in position, make sure one of your friends has a good hold of it – mirrors are expensive and come with seven years' bad luck built in. Carefully stand on the table parallel to the mirror and swing one of your legs over the top so it's hanging down on the reflective side of the mirror. Don't touch the floor. To the person watching it will appear as if you're floating above the floor.

How It Works

Although the person watching understands that your leg is being reflected in the mirror, their eyes trick them into thinking that the reflection is actually your other leg. That's why you appear to be floating.

70 ✎ MAKE A SOLAR BATTERY

Solar power is going to become increasingly important in the next 100 years. We've seen elsewhere how you can use the power of the sun to cook food in a solar oven (see page 60), but you can also use it to create energy, via a simple solar battery. While this experiment isn't going to power your house, it does demonstrate the essentially simple principles behind DIY solar power.

What You Need

Two sheets of copper flashing about 15cm (6in) square, a pair of electrical leads with alligator clips at both ends, table salt, tap water, some sandpaper and something to cut the copper sheeting with. You'll also need a heat source that glows red-hot, like an electric stove (gas is much more difficult to use), and a meter that can measure very small amounts of current (probably not much more than 20 or 30 microamperes); a modern multimeter will probably do the trick, but check before you buy it. Finally, you'll also need an empty two-litre plastic bottle that you can cut in half.

What To Do

Don't turn the cooker on yet. Cut the copper flashing to size and shape, then wash your hands thoroughly with detergent to get rid of any oil on them, and do the same to the copper. Also give the copper a good scrub with the sandpaper to remove any surface corrosion, before popping it onto the burner.

Turn the burner to high, sit back, and watch. As the copper flashing heats up, it undergoes some lovely changes in colour and it starts to oxidise. After a while, an ugly dark coat of cupric oxide replaces these colours, and by the time the burner is red-hot the copper will be almost entirely black. Keep an eye on it, but leave it on the burner for around 30 minutes so that the layer of cupric oxide thickens – this makes it easier to scrape it off.

Once half an hour has passed, turn off the burner and leave the copper where it is to cool naturally. It's important that you let it cool in its own time or you'll find the cupric oxide very hard to remove. This bit's interesting. The copper and the cupric oxide cool at different rates, and this mismatch causes black flakes to literally pop off the copper sheet. Leave the copper to cool for about 25 minutes. By now, most of the black will have flaked off by itself. To remove the rest, rub the surface gently under a running tap with your fingers. Now you should see the red cuprous oxide layer that you need to make your battery work.

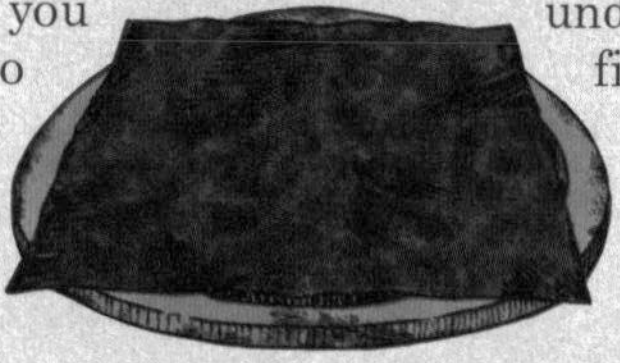

Fig. 87. After the copper turns black, leave it to cool for about 30 minutes. After that, most of the top will flake off to reveal a layer of red cupric oxide underneath.

Solar Power

With conventional fossil fuels running out and continuing general nervousness about nuclear power, the idea of capturing and using energy from the sun has never been more attractive. Gadget lovers can already charge their mobile phones and iPods using a solar charger, and many home owners are being encouraged to consider adding hollow panels to their roofs which are part of a water-pumping system. When the sun shines, the water is pumped to the panels where it heats and is then routed round the house and back to the water tank. This reduces the amount of electricity or gas used to heat your water. On a larger scale, various alternative technologies are being considered. One idea that might work well in very hot countries is what's being called a 'solar tower'. Essentially this is a tall tower in the centre of an enormous greenhouse. When the greenhouse is warmed by the sun, it produces hot air, which rises up the tower, driving turbines as it goes and producing electricity.

Next, measure up the plastic bottle just over 15cm (6in) from the bottom and cut it in half. Take the second sheet of copper (the one you didn't heat) and bend it so it wraps round inside the plastic bottle. Then – gently – do the same to the one you heated, so they're facing each other across the bottle. Make sure they don't touch. Take one of the alligator leads, clip it onto the unheated copper plate and hook up the other end to the positive terminal on your meter. Attach the other clip to the heated copper plate at one end and hook the other end up to the negative terminal on the meter.

Finally, mix a couple of tablespoons of salt into some hot water and then pour this into the bottle until it's about 2.5cm (1in) from the top of the plates. Make sure you don't get the clips or the leads wet. Put the whole contraption in the sunlight and watch what happens. On a sunny day you might see as much as 50 microamperes registering on the meter.

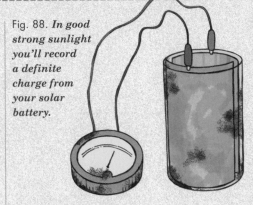

Fig. 88. *In good strong sunlight you'll record a definite charge from your solar battery.*

How It Works

The cuprous oxide that you revealed when the black cupric oxide layer rubbed off is a semiconductor. When it's placed in sunlight, some of its attendant electrons get sufficiently charged to break away into the salt water and then into the untreated copper plate. From here they travel down the electrical wire to the meter, where they're registered by the movement of the needle.

(71) BALLOONS THAT BLOW THEMSELVES UP

Feeling lazy, or do you just want to show your friends how you can blow up a balloon without actually blowing into it? This experiment shows how you can use air pressure to do the hard work for you.

What You Need
An ordinary glass bottle (a milk bottle will do the job), a balloon, a pair of scissors, something like a washing-up bowl, and some hot water (note that the water should be hot rather than boiling).

What To Do
Start by taking the balloon and using the scissors to cut off the narrow bit you'd normally blow into, and then put this to one side. Pour some hot water into the glass bottle, filling it nearly to the top, then leave it to stand until the glass itself become warm to the touch. When it's warm, pour the water away. Pull the balloon over the top of the bottle (depending on the size of the bottle and how well you cut the balloon, you may have to hold it there) and watch what happens. Without your doing anything, the balloon starts to inflate.

Now try exactly the same experiment, but this time have a bowl full of cold water handy. Put the bottle into the bowl and again hold the balloon over the top of the bottle. This time, instead of inflating, the balloon is actually sucked into the neck of the bottle.

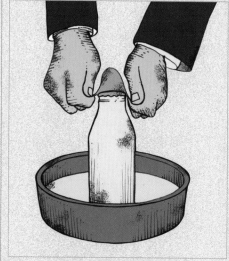

Fig. 89. *When the air in the bottle is warmed by hot water, it expands and will inflate a balloon held over the neck.*

How It Works
It's to do with air pressure again. When you stand the warm bottle on its own and put a balloon over the top, the air inside expands and so pushes out and up to blow the balloon up. When you put the bottle in cold water, the air inside contracts and takes up less space, which means that the air pressure outside the bottle is greater than that inside it – so the balloon is pushed down into the neck of the bottle.

Fig. 90. *If the water around the bottle is cold, the air contracts and sucks the balloon into the bottle.*

72 IN AND OUT AND IN AND OUT

If you've ever enjoyed making yourself dizzy, just to see what it felt like, then you'll like this one. It's going to mess – temporarily – with your visual cortex and produce some very interesting results.

What You Need
Just some cardboard, white paper, glue, a spiral and something to rotate the spiral on, like a record player; you can even poke a pencil through the middle and spin it, as in the experiment on page 64. Depending on how you choose to create the spiral, you'll also need a nail, a thick pencil, some thread, a ruler and a drawing compass.

What To Do
Start by sticking the paper on top of the cardboard with the glue. When it's dry, mark out the circle. You can use a compass if the circle's small, but if you want to make a bigger one, find the middle of the card and push the nail into it. Tie one end of the thread to the nail and the other to your pencil so you can draw out a larger circle using the nail as the centre of the 'wheel'.

To draw the spiral, wind the thread round the nail until the pencil is next to it. Then, holding the nail firmly, put the pencil to the paper and unwind the thread. As you do, you'll produce a spiral on the paper. This isn't easy. If you can't do it, try this alternative.

Draw a line across the middle of the circle with the ruler. Take the compass and draw a half circle above the line. Turn the paper round, open the compass out slightly and draw another half circle on the other side of the line. Turn the paper round again, open the compass a little more and draw another half circle. Repeat until you have your spiral. Poke a hole in the middle and put it on the record turntable and watch it spin for 20 seconds. Now look at anything nearby – it will expand or contract in a very unsettling way!

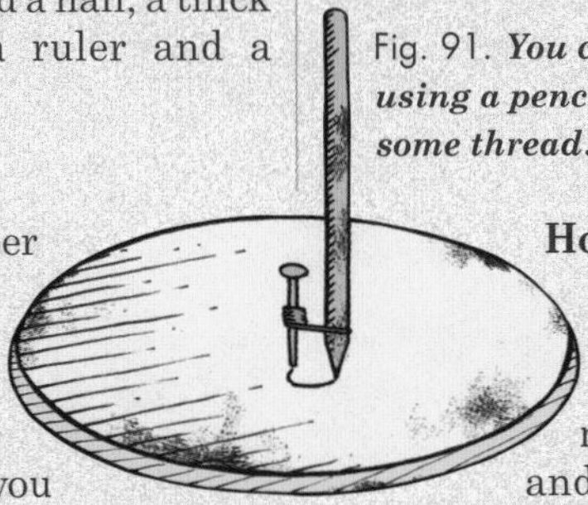

Fig. 91. *You can draw a spiral using a pencil, nail and some thread.*

How It Works
Parts of your visual 'nervous system' work harder when objects are moving away from you, and others work harder when things are moving towards you. When you look at the spiral, the one that detects objects moving away from you goes into overdrive, so that when you look away at something stationary, the one that hasn't been working is stronger than the tired one – so the object will appear to be moving towards you.

Fig. 92. *Place the spiral on a record turntable, start it spinning and watch what happens.*

73 TWO BOTTLE TRICKS

Here are two amusing tricks for the price of one! Both are easy to set up, and use a traditional style of glass bottle that has a stopper in the top.

Take the bottle off the heat, let the bubbling die down, and then call your friends. Tell them you can make the water in the bottle bubble and boil again by magic. To do this, blow through a straw (or put an ice cube) onto the surface of the bottle and the water will boil.

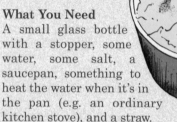

What You Need
A small glass bottle with a stopper, some water, some salt, a saucepan, something to heat the water when it's in the pan (e.g. an ordinary kitchen stove), and a straw.

Fig. 93. *The glass bottle – with stopper in place – being heated in salt water.*

What Do To
Fill the small glass bottle with water until it reaches about 2.5cm (1in) from the top, and then put the bottle (without the stopper in place) in a pan of salt water.

Heat the pan until the water in the bottle is boiling.

Take the pan off the heat, allow the water vapour time to drive out any air in the bottle, and then put the stopper on very firmly. As the bottle cools, the water vapour will condense, leaving a vacuum inside.

Now for the first trick, tip the bottle up and right it again. As the water pounds about in the vacuum it will sound like the banging of a hammer!

For the second trick, simply put the bottle back in the pan of water and boil it again.

How It Works
The key to this trick is pressure, and the effect that it has on boiling points. We all know that the boiling point of water is 100°C (212°F), but this is not always the case. Pressure can affect the exact temperature at which water boils, which is why it takes less time to boil a kettle on top of a mountain than it does at sea level. The lower the pressure, the lower the boiling point; the higher the pressure, the higher the boiling point. You are able to make the water appear to 're-boil' because, by blowing on the bottle, you have cooled it, making the water vapour condense and reducing the pressure inside. As the pressure decreases, the heat left in the water is enough to make it boil again.

74 HEAVY AIR

Air is weird. You can't see it, most of the time you can't feel it, and yet it's all around us and as you read this is actually pushing down on you with a force of 15lb per square inch. No wonder you're always tired.

What To Do

For this experiment you'll need a sturdy table, a thin piece of wood about 60cm (2ft) long by about 5cm (2in) wide and 3mm (^{1}/8in) thick, a couple of large sheets of newspaper and a hammer. Lay the wood on the table so that about 10cm (4in) is sticking out over the edge. Cover the rest of the wood with the two sheets of newspaper, making sure that they make a nice newspaper–wood–table 'sandwich'. Take the hammer and whack the end of the wood really hard. Instead of flipping up and off the table, the piece of wood that's sticking out snaps clean off.

How It Works

If you hit the wood hard and fast, the air can't get underneath the newspaper fast enough to equalise the air pressure above the newspaper. Thus, the paper is able to hold the rest of the wood in place and the smaller end piece snaps off.

Hit the end of the wood sharply once with a hammer

75 THE MAGNETIC BOAT

A lot of the experiments you can do to show the effect of magnetism are worthy but dull. This one is much more fun, and you end up with something fun for the kids as well.

What To Do

For this you need a magnet, an offcut of wood, a small handsaw, some sandpaper, a couple of nails, a couple of horseshoe-shaped tacks, a hammer, a piece of paper, a large aluminium pie dish, and four similarly sized blocks of wood. Saw one end of the wood offcut to make a rough bow, and then sand it smooth. Next, lay one nail along the bottom of the 'boat' and hammer the tacks at each end to secure the nail. Poke a couple of holes in the paper to make a sail and then thread the other nail through the holes and hammer it into the top of the boat. Put the tray on the blocks of wood, sit the boat on the tray and then 'sail' the boat with the magnet.

How It Works

The nail underneath the boat is attracted to the magnet, so as you move it around, the boat follows.

Young kids will enjoy this simple experiment

76 THE CABBAGE CHAMELEON

A red cabbage is a red cabbage, right? Wrong. You'll be amazed at the way this innocent vegetable can be encouraged to take on different colours.

What You Need

Some red cabbage leaves, a vegetable grater, a bowl of cold water, a colander, some clear vinegar, baking soda and some lemon juice (either from a real lemon or out of a bottle, it doesn't matter). You'll also need some plastic cups and a jug. If you want to really push the boat out, try all sorts of other household items like soap or detergent and see what happens when you add them to the cabbage water.

What To Do

Grate the cabbage into the bowl and add the cold water. Leave it to stand for an hour and then strain it through the colander into the jug so you've got a ready supply of rich, purple-coloured juice. Pour the same amount of juice into three plastic cups, then add a little baking soda and watch what happens. The mixture will turn a strong green colour. Next, add some of the lemon juice to the second cup and watch as it turns bright pink. Add the vinegar to the third cup and watch what happens.

How It Works

It's all to do with the pH of cabbage water. This is simply a way of measuring whether something is acid or alkaline. When you soak the cabbage shreds in water you extract things called anthocyanins. One of the peculiarities of these molecules is that they change colour depending on the relative pH of the environment they're in. So, a pink result means the mixture is highly acidic, while a green result means it's alkaline.

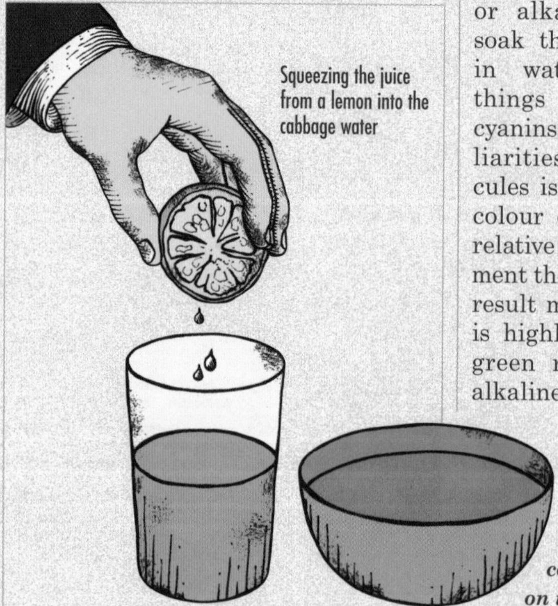

Squeezing the juice from a lemon into the cabbage water

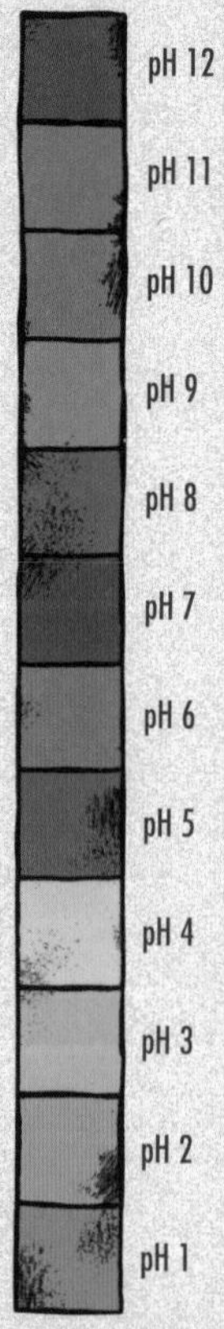

Fig. 94. *Different substances will react with the water to produce different colours, depending on whether they're acid or alkaline.*

pH 12
pH 11
pH 10
pH 9
pH 8
pH 7
pH 6
pH 5
pH 4
pH 3
pH 2
pH 1

77 A COMPRESSED-AIR ROCKET

There are several rockets in this book, but this compressed-air version is easily the most powerful. Rockets launched like this can travel hundreds of feet.

What You Need

A plastic bottle that's had fizzy drink in it (it'll be stronger than one that's been used for still drinks). You'll also need something to use as a bung for the bottle – a cork is an obvious choice, possibly one of the plastic ones that are used for wine bottles. Whatever you choose, it has to be a good fit. Then you need a bicycle pump, something to use as a launch pad (perhaps a couple of logs for the fire or some bricks) and the type of adaptor you use in order to inflate a football (basically a sharp nozzle that screws onto the pump).

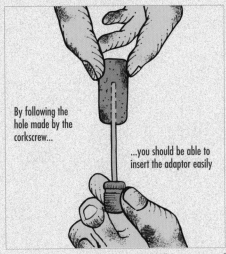

By following the hole made by the corkscrew...

...you should be able to insert the adaptor easily

What To Do

First you need to make a hole in the cork and jam the football adaptor into it. Fortunately, this needn't require you to heat a nail until it's red-hot before poking it through, because your cork should already have the beginnings of a hole, courtesy of the corkscrew. So, push and twist the adaptor into the cork and then push the cork into the bottle. Attach the pump and give it a few goes to get some air into the bottle – you need to do this to make sure there aren't any holes in the bottle that you haven't spotted, and that the cork and football adaptor contraption fits snugly. Remove the cork and fill the bottle to halfway with water. Jam the cork back in. Rest the bottle on the launching pad and make sure that it's pointing away

from people and property (because of the way the adaptor is positioned, it's hard to make the rocket take off vertically, so position it at an angle). Connect the pump to the adaptor and start pumping. After a short while, your rocket will take off.

How It Works

As air builds up in the bottle, the pressure forces the cork out. Newton's third law of motion (for every action there is an equal and opposite reaction) now takes over and as the water rushes out of the back, the rocket is propelled forward.

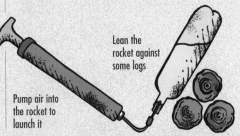

Lean the rocket against some logs

Pump air into the rocket to launch it

(78) MAKE A PINHOLE CAMERA

In the age of digital photography it's hard to remember that you can take photographs with a home-made camera that costs almost nothing. Instead of forking out for a digital camera and computer, all you need is a few bits and pieces and the price of your film and development.

Fig. 96. *When you're ready to take your photograph, you can peel back the sticking plaster to reveal the pinhole.*

What You Need
This really does use the simplest of materials. You'll need a shoebox, but one that's in good condition, closes properly and is what photographers call 'light-tight' – in other words, it can keep out most if not all of the light. Then you'll need some black paper, some tracing paper, sticky tape, glue, a pin, a sticking plaster, and a pair of scissors. You'll also need some film for your camera – more of which later.

Fig. 95. *Use a pin to make a tiny hole in the box end – the smaller the better.*

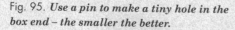

What To Do
First, take your shoebox and line it with the black paper, gluing it to the sides and bottom and also to the lid, remembering that you're trying to make the box as film-friendly as possible. Next, take the scissors, cut a rectangular hole in the centre of one end of the box and cover it with tracing paper, secured with tape. Take the box, put it on its end (tracing-paper side up), and use your pin to 'pop' a tiny hole in the centre of the tracing paper; the smaller the hole, the better your pictures will be. There are different ways to make a shutter, but one of the simplest is to use a piece of sticking plaster where the soft pad covers the pinhole. Stick the plaster to the front of your camera, as shown in the illustration.

Which brings us to the question of what you're going to use for your actual photos. There are several options, each with its own pros and cons. You could use sheet film, but this will produce a negative; you could use something called liquid emulsion, which is painted onto paper to make it light-sensitive, but this is a bit hit-and-miss and still needs developing. Probably the best bet is simple black-and-white print paper, which you can buy in various standard sizes – for example, 7x5in or 5x4in – and then either develop yourself or take to a specialist photo lab.

Whatever film you use, it's important that you don't expose it to light either when loading your camera or when you take it out, so you'll want a dark room and/or a thick black-cloth bag that is big enough to take the camera and let you load the film direct from its packet (photo shops sell things called 'changing bags' that have holes for both your arms). Follow the instructions on the film about which side should be facing

Fig. 98. *Start by taking a picture of something on a window sill to ensure that there's plenty of light.*

the pinhole, and fix it in place with a small piece of tape top and bottom. Once the lid of the camera has been replaced, pop the camera on something facing the window so there's plenty of light, and peel the shutter back. Leave it for 15 minutes, and then put the shutter back on. Put everything back in the bag, remove the film and keep it somewhere safe, away from the light, until you get it developed.

How It Works
The human eye is able to see because rays of light, reflected by objects, are focused onto the retina. The light-sensitive film in a pinhole camera performs exactly the same function as the retina, except it records the image permanently. Now, because individual points of light are the building blocks of photos, it stands to reason that if the points are too large, the image will be fuzzy and unfocused. That's why it's so important that your pinhole really is the size of a pin.

Fig. 97. *You need to be careful and not expose your photographic paper to natural light, or it'll be ruined.*

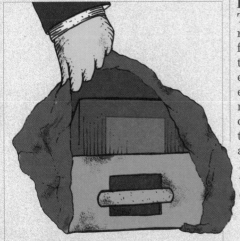

79 | THE CAN CRUSHER

Here's something that demonstrates a sound scientific principle, is plenty of fun, and at the end leaves an innocent can crushed on the floor. And no, you don't have to stamp on it. Stand back – it's violence meets science!

What You Need

A large metal can with a decent screw top that can be used to produce a sound seal. The top is the key component to this experiment, because if you can't make the can airtight, it won't crush. Good cans include those that have contained a liquid that needs to be held securely, such as oil. However, that means you also need to make sure that they're emptied and cleaned thoroughly before going ahead with this experiment. A strong household detergent should clear up any traces, while sluicing out with plenty of cold water will finish the job. You'll also need a little bit of tap water, a pair of oven gloves or a thick cloth, and something to heat the water and can to boiling point.

What To Do

Pour about 13mm (1/2in) of water into the can – this doesn't have to be too precise, you just need enough water to boil and make some steam. With the top still off, heat the can until the water starts to boil, then take it off the heat. As soon as the steam starts to die down, remove the can from the heat with the gloves or cloth and screw the top back on tightly. After a few moments the can will start to buckle as if being crushed by giant invisible hands.

How It Works

As the can starts to cool, the steam that's left inside condenses and the result is a partial vacuum. That means the pressure of the air outside the can is much greater than the pressure inside, and so the can is crushed by the power of the air outside.

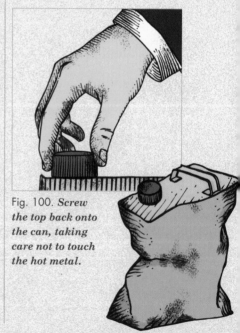

Fig. 99. *Make sure the lid is off the top of the can before you start to heat the water inside it.*

Fig. 100. *Screw the top back onto the can, taking care not to touch the hot metal.*

80 MEASURING SURFACE TENSION

Like any liquid, water has an invisible 'skin' – created through surface tension – which some insects, for example, use to scuttle across the surface. To see it in action, all you need for this experiment is a glass, a bowl, some water, an eye dropper, a piece of tissue paper and a small needle.

What To Do

Start by filling the glass almost to the brim, and then continue adding water with the dropper. As you do, you'll see that the level of the water actually bulges out over the top of the glass without spilling – so long as nobody knocks the glass. Alternatively, try this. Tear off a small square of tissue paper and put a pin on it. Pop them both gently into the water and watch what happens. After a short while the paper will become soaked and sink to the bottom of the glass. The pin, however, will continue to float on the surface.

How It Works

The surface of your glass of water has more water on one side and air on the other. This makes the molecules stick more closely together and provides the necessary tension so that the pin doesn't sink.

Fig. 101. *Amazingly, the needle will actually float on top of the water, thanks to surface tension*

81 LIGHT UP A FLUORESCENT BULB

We've already done some simple experiments with balloons to demonstrate the power of static electricity, but here you'll discover how you can use it to make a light work. All you need is a fluorescent light bulb and a dark room.

What To Do

Blow the balloon up and then turn the light off. Hold the bulb in one hand and with the other, rub the balloon on your head for 30 seconds (it'll help if you have a full head of hair). Bring the balloon and the bulb together and watch what happens – the bulb will flicker into life. In fact, if you rub the balloon against your head again and then move the bulb up and down near the balloon, the light will appear to follow the balloon.

How It Works

Static electricity. Electrons from your hair jump over to the balloon and when you bring them near the bulb they stir up the mercury vapour that's inside the fluorescent tube. This sends out ultraviolet light, which in turn hits the phosphor coat on the bulb and lights it up.

82 MAKE A WATER-POWERED FUEL CELL

The experiment on page 17 demonstrated how to 'split' water into its component parts; this experiment takes that several stages further so that you end up with a glass of water that can produce electricity. Sounds impossible? Read on and find out all about the water-powered fuel cell.

What You Need

Probably less than you think. In fact, you only need a couple of specialised bits of equipment. First, you need to track down about 30cm (12in) of platinum-coated wire – an old-fashioned jeweller's shop is probably the best bet for this. Second, you need a multimeter for measuring the electricity that your water-based cell is going to be producing. After that, it's pretty straightforward: a pair of pliers, an old lollipop stick, a 9-volt battery, some transparent tape, a 9-volt battery clip (the two wires and the little plastic bit with the two terminals on, sometimes called a battery snap) and a glass of water.

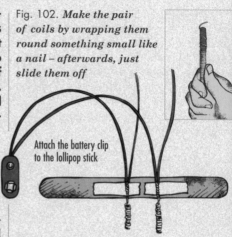

Fig. 102. *Make the pair of coils by wrapping them round something small like a nail – afterwards, just slide them off*

Attach the battery clip to the lollipop stick

What To Do

Start by using the pliers to snip the platinum-coated wire in half to make two 15cm (6in) lengths; then find something to wrap the wire around in order to form two tight, coiled cones (a small nail works well). These will act as the electrodes in your fuel cell. When you're done, the cone should like the one in Fig. 102, with a little 'tail' sticking out. Repeat for the second electrode.

Next, take the 9-volt battery clip and have a look at the two red and black wires coming out of it. Separate the wires and cut them in half so you've still a red and black wire coming straight from the battery clip and also two 'spare' red and black wires. Remove a little of the insulation from the free ends of all four wires with the pliers. Now you need to attach one end of both red wires to one electrode and one end of both black wires to the other. Twist the wires carefully round the

Hydrogen Fuel Cells

The possible application of fuel-cell technology is one of the most exciting branches of modern science. Fossil fuels are bad for the environment in many ways: from extracting them in the first place, right through to the emissions they give out when we're using them. A hydrogen fuel cell, on the other hand, produces only electricity and water – it doesn't burn anything, is exceptionally quiet and doesn't cause any pollution. A hydrogen fuel cell fitted to an electric car, for example, would take about five minutes to refuel and deliver the same mileage as conventional fossil fuel, but with all those added benefits. Many automobile manufacturers have produced working cars that use fuel-cell technology. It's not going to happen tomorrow, but most scientists are convinced that it's only a matter of time before fuel cells change the face of motoring. Opponents point to the expense of hydrogen fuel cells, but they're only considering the prototypes and small-run vehicles that are currently available – mass-produced cars would be much more competitive. Think back to what a car (or indeed a computer, another example of a 'new' technology) used to cost and it'll give you an indication of how far prices will fall once the vehicles go into full-scale production. Projections looking at cost per mile already indicate that cars using hydrogen fuel cells will actually work out cheaper to own and to run.

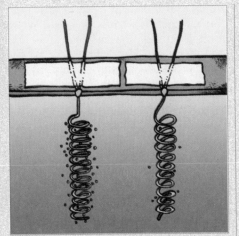

Fig. 103. *As the current passes through the fuel cell, hydrogen and oxygen bubbles gather round the electrodes*

electrodes and then secure them with the transparent tape. Then stick both wires to the lollipop stick about 20mm (³/₄in) apart.

Next, take a glass and fill it almost to the top with water. Position the contraption you've just made so that most of the electrodes are submerged, taking care not to get any of the wire connections wet. Then take your multimeter and connect the spare black wire to the negative terminal and the red wire to the positive one. Amazingly, your water-powered fuel cell is complete. All you have to do is give it a kick-start.

Take the 9-volt battery and just touch it to the connectors on the battery clip (you don't need to plug it in because we only need it to provide current for a couple of seconds). As you do, you'll see that the multimeter registers the

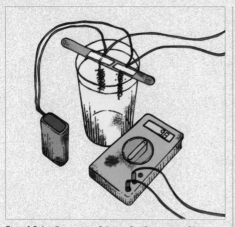

Fig. 104. *Just touching the battery clip to the points on the top of the battery is enough to send a charge into the cell and start the process.*

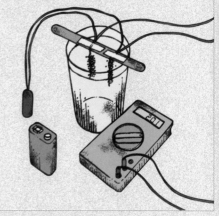

Fig. 105. *Even without the battery, the reaction inside the glass should be enough to generate a small amount of electrical current.*

current passing down the wires and jumping between the electrodes. However, keep watching and take the battery away. You'd expect the current to drop to zero, but instead, although it drops sharply from the 9 volts being generated a second ago when you held the battery to the clip, there's still some voltage being produced. This time, the glass of water is producing it. You've made a water-powered fuel cell.

How It Works

First of all, when you touch the battery to the battery clip and send current through the wires, this makes the water around the electrodes split into hydrogen and oxygen, courtesy of electrolysis (see page 17). As then, you can see the bubbles hanging onto the ends of the electrodes. What's different about this experiment, though, is what happens next. Remember that the electrodes are made up of platinum wire? This acts as a catalyst and lets the oxygen and

hydrogen re-combine. When they do, something interesting happens.

Without the battery passing current, the hydrogen bubbles around the electrode break up, courtesy of the platinum, which is acting as a catalyst to form positively charged hydrogen ions and electrons. Meanwhile, over on the other electrode, the oxygen bubbles pull electrons from the metal and then combine with the hydrogen ions. Result? Water. And the byproduct of this reaction? Electricity. Not very much, but enough to register on the multimeter. In fact, you'll probably be producing about as much electricity as William Robert Grove did way back in 1839 when he first conducted this very experiment and created the first fuel cell. At least your fuel cell is completely clean and environmentally friendly – unlike Grove's original, which produced an unpleasant side effect when it spewed out poisonous nitric oxide gas.

83 HOW TO MAKE BUTTER

You don't have to work on a farm to make butter; this experiment shows you how to turn your shed into a dairy. All you need for this is some double cream and a decent-sized jar with a lid.

What To Do
Open the jar and pour in the cream. Close the lid and the begin shaking the jar as if you were playing percussion in a samba band. It may take some time before anything happens – as long as 10 minutes – so you'll need to be patient. After that you'll discover that the cream inside the jar has changed in composition and that in fact it's divided into something more solid (butter), with a liquid residue – what the Americans call buttermilk.

How It Works
The two main components in the cream are proteins and fats. By shaking the jar you get these to cling together, and the more they cling together, the more proteins and fats they attract, a bit like a snowball rolling down a mountain. The buttermilk that's left over is mainly used for cooking.

84 MEASURING WIND SPEED

The next time you get a windy day, take this experiment out into the garden and see how easy it is to see wind power in action.

What To Do
For this you'll need: four plastic or paper party cups, some paints and a brush, a stapler, a ruler, a large drawing pin, scissors, a pencil with a rubber on the end, two strips of cardboard and a watch that can display seconds. Paint one of the cups a bright colour and leave it dry. Staple the cardboard strips together to make a cross, then staple the cups onto the ends of each arm – making sure the cups all point in the same direction. Then take the drawing pin and push it through the exact centre of the cross, and take the rubber end of the pencil and push it onto the pin. This should hold it securely, yet allow the anemometer to move freely as you hold the pencil.

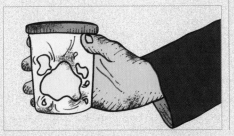

Make sure the cups are all pointing in the same direction!

How It Works
As the wind blows it meets resistance from the cups and pushes them round. Using the second hand of your watch, and the coloured cup as a marker, you can roughly judge the speed of the wind.

101

85 FIREPROOFING

Like most of the experiments in the book, this has its basis in sound science, but you can also use it to trick your friends into thinking you've managed to fireproof an ordinary piece of cotton thread.

What You Need
Some cotton thread, a small bowl, some household salt, something metallic but light – like a paperclip or a curtain ring – a ruler to hang it from, and a box of matches.

What To Do
Like many things, the trick is in the preparation. Add salt to a bowl of water until you can't dissolve any more, then pop in the thread. You can actually prepare several threads at a time so you can play the trick on different people. Leave the thread in the bowl for an hour, take it out and dry it off, then put it back in the bowl and soak for a further two hours. Repeat once more, and the thread is ready for action. Tie one end to the ruler and tie the paperclip to the other end. Slide the other end of the ruler under a heavy object on a shelf so that it's held securely and leaves both of your hands free to light the match. Show your audience that you have an ordinary piece of cotton thread (it will certainly look ordinary, despite the repeated salt soakings) and then set fire to it. It will seem to your friends as if you are unable to burn the thread.

How It Works
Of course the thread does burn, but the column of salt which has woven itself through the fibres does not, and is strong enough to keep the paperclip (or whatever you ended up using) suspended at the end. That's why it's important that you don't try to hold the ruler – the slightest movement could break the column of salt – and that the object at the end of the thread is light.

Fig. 106. *Although the string burns away, the column of salt remains in place and should be strong enough to support the weight of the paperclip.*

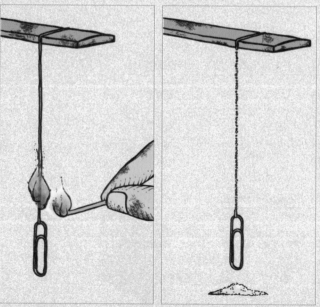

86 BUBBLE POWER

Bubbles might seem like a harmless bit of fun, which they are, of course. But they also combine strength with flexibility and demonstrate some remarkable properties to do with surface tension.

What You Need
An empty plastic bottle of washing-up liquid, a sharp knife, two straws and some water, plus a candle and something to light it with.

What To Do
Start by making a 'pipe' big enough to blow some large bubbles. The easiest way to do this is to measure off (from the top) about a quarter of your washing-up-liquid bottle and then cut across it with a sharp knife. Put the rest of the bottle to one side (it may be useful for other experiments elsewhere in the book) and turn the end you've just cut off upside down. That's the bowl of your pipe.

Use the tip of the knife to poke a hole in the side and twist in the straw. Add a little water to the bowl and give it a stir – even though it's 'empty', there'll be enough washing-up liquid left to produce some good bubbles.

Next, light a candle, blow a big bubble, and catch it carefully on the end of the second straw. Then cover the other end of the straw with your finger. Now move the bubble towards the candle flame. Nothing. Take your finger away from the end and watch what happens. As the bubble nears the candle, the flame shies away from it as if being pushed by an invisible force. Meantime, the bubble shrinks and will eventually disappear.

By the way, if you're having problems making a mixture that produces really good bubbles with plenty of staying power, then try adding either some glycerine (usually available from a chemist or supermarket) or some corn syrup. Start with a tablespoon and add more if you need to.

How It Works
When you take your finger away from the end of the straw, the particles in the bubble contract to form little droplets, and the surface tension pushes the air out. The bubble is so strong that it pushes air away from its surface and thus appears to blow the candle flame.

Fig. 107. *When you remove your finger from the other end of the straw and continue to inch the bubble towards the flame, you'll see it 'flinch'.*

Washing-up liquid will produce good bubbles

The flame will actually move away from the bubble

87 MAKING SPARKS FLY

There are lots of experiments that use the power of static electricity to make them work, but sometimes you get the feeling you're not really seeing the real thing in action. That's not the case with this – you're really going to see sparks fly!

What You Need
A disposable aluminium pan (the type that pies come in), a pencil (the kind with a rubber on the end), a drawing pin, a styrofoam plate and some wool.

What To Do
Get the aluminium pan, turn it over and press the drawing pin through the middle. Then turn it back over again so the inside of the plate is facing the ceiling, with the pin sticking out. Take the pencil and press the rubber end firmly onto the pin – it needs to be secure enough for you to pick up the plate using the pencil. Give the styrofoam plate a really good two-minute rub with the wool and then put the plate upside down on a flat surface. Pick the aluminium plate up using the pencil and put it on top of the styrofoam plate. If you touch the aluminium pan with your finger, you'll get a small shock, but that's not when the sparks really fly. Take the metal plate off again and then touch it with your finger. Ouch. Try it in a darkened room and you'll see those sparks.

How It works
Essentially, you've built something called an electrophorus – a simple generator that can make static electricity. Rubbing the styrofoam plate against the wool attracts electrons and produces a negative charge. When you put it on the metal plate it repels the electrons there, but since they can't go anywhere, the metal pan stays neutral. If you touch it while the styrofoam plate is still there, the electrons jump off the pan and onto you, creating a spark as they jump through the air to you. Now the pan is charged positively (courtesy of induction) and if you go to touch it again, you'll create a second spark.

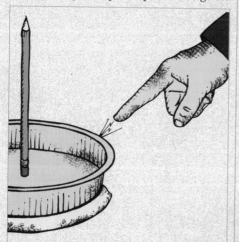

Fig. 108. *This experiment generates enough static electricity to give your finger a good 'ping' – you may even see a spark if you do it in a dark room.*

88 BUILD A HOVERCRAFT

Elsewhere in this book you'll discover how to build a small balloon-powered hovercraft (see page 42) that will teach you the basics of design, aerodynamics and propulsion. This larger one, however, is big enough for you to sit on and would make an excellent project for a group of friends.

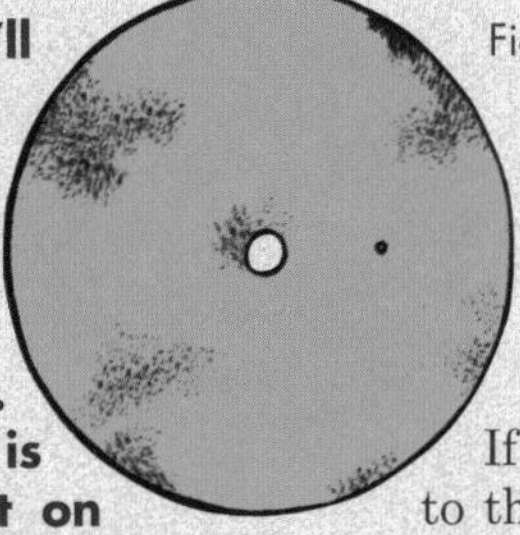

Fig. 109. *Plywood makes a good base for the hovercraft because it's both light and reasonably strong.*

What You Need

Necessarily, this experiment needs a little specialised equipment, but it's the sort of item that more and more homes are investing in – and anyway, if you're serious about hovercraftery, then you can always club together with some friends and hire one to get you started. The main item is a leaf blower – and preferably a battery-powered one so you can move it about with the hovercraft.

Fig. 110. *Something like a leaf blower is excellent for generating the kind of propulsion that will get your hovercraft on the move.*

If you use one that's tethered to the mains supply, it's going to restrict your hover-related fun to a certain extent. If you don't have a leaf blower and have no way of getting one, you can try to track down an old-fashioned vacuum cleaner that has a blower feature – it won't generate as much puff, though.

Moving on, you'll need a good bit of plywood of about 1.2m (4ft) square and 1cm (3/8in) thick. If your intended passengers are larger than average, you may want to think about increasing the thickness of the plywood. However, you don't want to make it too thick or it'll impede the hovercraft's performance. You'll also need a strong plastic sheet about 30cm (1ft) larger all round than the plywood base. This needs to be heavy-duty plastic – think shower curtain rather than bin bag. You'll need to scrounge the lid from one of those large catering-sized coffee tins that people buy in offices, get a large bolt – about 5cm (2in) long and 8mm (5/16in) across – with a nut and a couple of washers to match. Find one of those extendable metal tape measures, some string, a nail and a pencil, and you're done. In terms of tools, you need a jigsaw, a drill, a hole-saw fitting for the drill, a sharp knife, some heavy-duty duct tape, sandpaper, and a staple gun.

What To Do

Start by cutting the corners off your square of plywood with the saw. The easiest way to do this is to find the centre of the wood first. Measure off halfway along both opposite sides and draw a faint line across the wood, and then do the same with the other two sides. The centre is where the lines cross. Tap the nail into the middle to make a notch, then cut off slightly over 60cm (2ft) of string. Tie one end to the nail and the other end to a pencil. Hammer the nail into the middle of the wood and then, holding the string taut, draw a circle all the way round. Use that as your marker when you're sawing.

Take your nail out and drill a hole through the centre of the plywood; this should be big enough to take the bolt. Next, take the round business end of your leaf blower and place it flush against the plywood disc, halfway between the centre and the edge. Draw around it with the pencil and use the drill with a hole-saw bit to cut out the circle. Clean up the hole and the edges of the platform with sandpaper.

Now you need to spread out the plastic sheet and position the wooden base on top of it. This bit's quite tricky. Fold the edges of the plastic up and over the edge of the board and then staple it to the top – about every 7.5–10cm (3–4in) should do. Try to think ahead to when the plastic is inflated to gauge exactly how tight to pull it; you're going to want the plastic under the board to puff out to about 10cm (4in) when fully inflated.

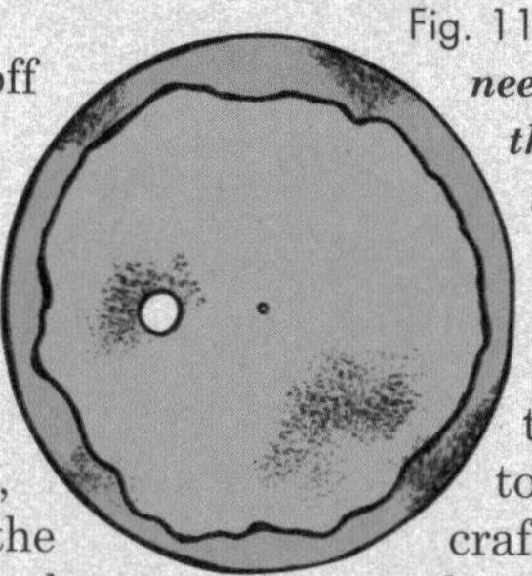

Fig. 111. *Your plastic sheeting needs to be pretty sturdy – think plastic tarpaulin rather than plastic bin bag.*

Next, drill a hole through the coffee-tin lid big enough to take the bolt, and attach it to the underside of the hovercraft as in Fig. 113 – bolt through washer, through lid, through plastic, through platform, out the top, through washer and secured by the nut. You've now got the basis for the classic doughnut shape that the hovercraft will take on when it's inflated. To finish the job, you need to cut a series of 5cm (2in) holes in the plastic sheet around the edge of the coffee lid. These need to be far enough apart so the plastic doesn't rip and close enough to the lid that they don't get clogged when the craft is at rest. Reinforce the holes by adding a little duct tape.

Turn the hovercraft over and insert the business end of the leaf blower into the hole you made in the top. Secure it with duct tape if it's not a tight enough fit. Switch on the blower and the hovercraft will rise from the ground. Give it a gentle push and you're away.

How It Works

When the air from the leaf blower inflates the plastic it pushes the entire thing up from the ground. The holes are positioned in and around the coffee-can lid (which is bolted

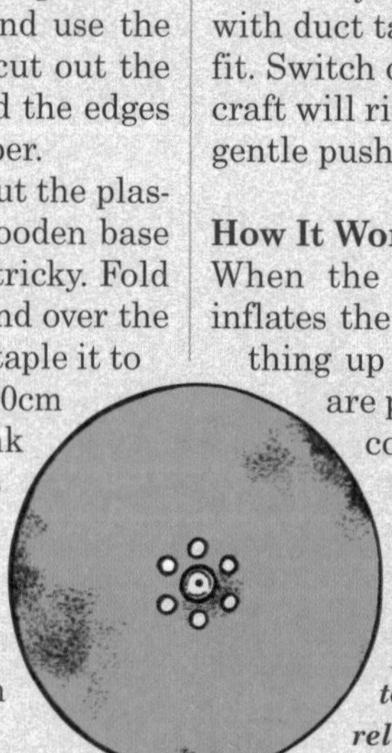

Fig. 112. *Seen from underneath, here's the approximate position of the holes you need to make in the plastic sheet, relative to the coffee-tin lid.*

to the platform and thus clear of the ground) and so when the plastic inflates, these blow the air out and pressurise the centre of the ring. As the air spreads out from the centre along the space between the plastic and the ground it creates a large area of low friction on which the hovercraft can glide.

The hovercraft as described here is a pretty big project – one of the largest in the book – but it doesn't have to be that way. If you prefer, you can build a smaller model with all of the dimensions suitably scaled down to hover unaccompanied, or perhaps attached to a tether so you can control it. A smaller craft will get round the potential problem of the leaf blower and allow you to use a vacuum cleaner that has a blow function. If you like the idea of a hovercraft but can't face one this big, turn to page 42 where you'll find details of the balloon hovercraft model.

Fig. 113. *The main components of the hovercraft in an 'exploded' view – you'll have to cut the plastic sheeting to size.*

Be careful not to rip the plastic sheet

Fig. 114. *Here the same ingredients are shown but with the sheet cut to size and attached to the plywood disc.*

You can 'finish' the hovercraft with duct tape if you like

The First Hovercraft

REAL SCIENCE

When the hovercraft was first invented in 1959 by Christopher Cockerell, an English boatbuilder, it was called by some a 'man made flying saucer' – that's how unique a vehicle it was. In a way, the description was true, since the hovercraft floated on a cushion of air, thus reducing the amount of friction between the craft and the surface of the water.

Cockerell's prototype was much less sophisticated than the one we describe here – he actually used a cat-food tin inside a dog-food tin and then produced a hover effect with an industrial air blower. The British government was so impressed by the possibilities offered by this new means of transport that it immediately placed Cockerell's hovercraft on the secret list – which meant, of course, that he couldn't develop it!

It took four years of hard work to convince the Ministry of Defence to release the plans and the first working model was built in Cowes on the Isle of Wight by Saunders Roe, a company that had made its name with flying boats (planes designed to land on water).

89 COLLECT MINIATURE METEORITES

Meteorite showers at night are wonderful things, but unlike the movies, you're unlikely to discover one (hollow, of course, and complete with alien) the following day in your back garden. Or so you might think. Actually, you can probably find the tiny remnants of miniature meteorites after every shower. You just have to know where to look.

What You Need

This experiment actually requires some proper scientific laboratory-style equipment in order for you to see these tiny meteorites, so the first thing you'll need is a microscope and some slides. After that, you'll need a couple of heat-proof glass dishes (the kind you can put in the oven or on a hotplate), a magnet, a small (i.e. slightly larger than the magnet) plastic bag, something to heat one of the dishes with, some distilled water and a sewing needle.

Fig. 116. *Make sure that no water can get into the plastic bag while you're moving the magnet around inside the bowl.*

What To Do

Take one of the dishes and give it a really good clean. Then clean it again. Then ask a friend if it looks clean and no matter what they say, clean it again. Then put it out in the garden and wait for rain. While that's happening, take the second dish, clean it to the same standard, and fill it with tap water. Put it aside somewhere safe to stand for a day or so – this will allow the chemicals commonly found in tap water (chlorine and fluoride) to evaporate. It's not exactly distilled water, but it will have the desired effect.

When the dish outside is nearly full of rainwater, bring it indoors. Take your magnet and pop it inside the plastic bag. Twirl the bag shut and then, keeping a firm grip on it so that no water can get

Fig. 115. *Not all meteorites are large enough to light up the night sky – many fall to earth unnoticed.*

in, run the magnet like a vacuum cleaner around the dish, under the water, making sure that you sweep the sides as well as the bottom. Carefully take the magnet and bag out of the first dish and pop them into the second dish, remove the magnet and then swish the bag around a bit. This will remove any tiny bits and pieces that the magnet's picked up.

Next, put the dish either into the oven or onto a hotplate and evaporate the water – for safety, remember to keep an eye on things. When the water's completely evaporated, take the needle and rub it across the magnet in the same direction for a couple of minutes so it becomes magnetised. When you've done that, rub the needle across the bottom and sides of the dish to pick up any tiny metallic fragments and then tap the needle onto one of the microscope slides. Look carefully through the microscope.

*Fig. 117. **When you rub a needle across a magnet for a couple of minutes in the same direction, it'll become magnetised and you can use it in this experiment.***

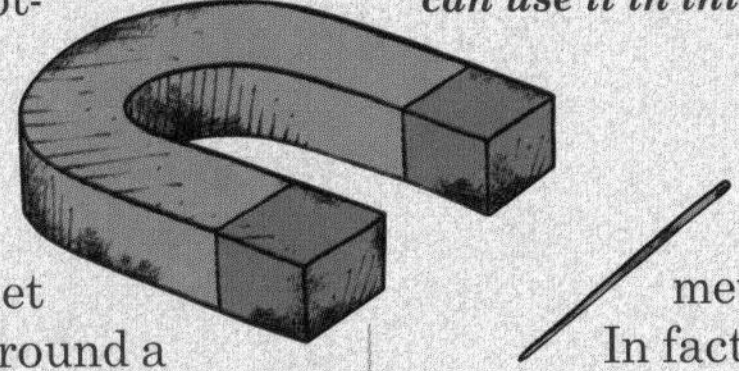

You should now be able to see your miniature meteorites. OK, not quite. In fact, most of what you'll see on the slide is likely to be almost anything but mini-meteorites – it's just as likely to be pollen or dust or any other gunk from the garden. However, if you can find rounded metallic particles, there's a good chance that they originated not on earth but somewhere in outer space.

How It Works

Many meteorites (those that have made it through the atmosphere) contain considerable amounts of iron and nickel, both of which are highly magnetic. Ignore any jagged shapes you may see under the microscope and instead look for smoother, round objects, possibly with a pitted surface. These are quite likely to be miniature meteorites that have survived their journey across the universe. Their diminutive size is one of the reasons they don't burn up in the earth's atmosphere, but rather drift around until they get picked up by much larger (to them at any rate) raindrops or hitch a ride with ordinary dust particles.

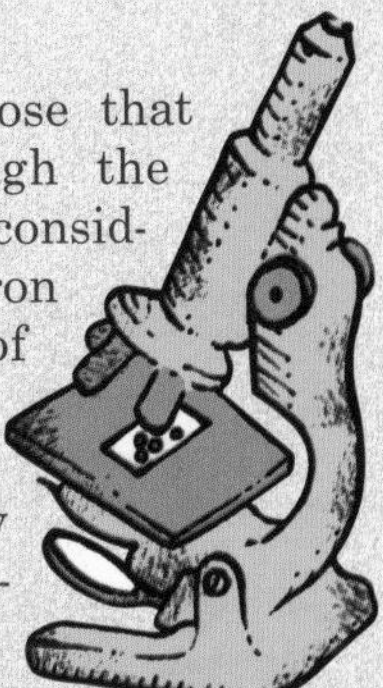

Check in the microscope for rounded metallic particles

*Fig. 118. **When the water's evaporated, gently rub the magnetised needle across the surface of the bowl to pick up any metallic fragments.***

90 THE BOOMERANG CAN

We're all familiar with the curious Australian stick that keeps coming back when you throw it, but did you know you can achieve the same results with a tin can?

What To Do

For this you'll need a tin of some description that has a lid and a bottom, a drill, some elastic, some thread, and a weight (a fishing weight will do the trick). Make a couple of holes in the top and bottom of the tin – they should be about 4cm (1½in) apart and arranged so they line up. Take the lid off and then thread the elastic through the holes so it goes in through the lid, out through the bottom, back into the tin through the bottom and out again through the lid again. Next, tie the lengths of elastic together with one end of the thread, and hang the weight on the other end. Put the lid back on, pull the elastic tight and tie it off. Roll the tin away from you and watch what happens.

How It Works

As the tin rolls away, the weight winds up the elastic until it reaches the point at which the tension (and energy) in the elastic are stronger than the momentum of the tin, whereupon it stops and then rolls back towards you.

91 DATE WITH DENSITY

This experiment shows how you can fill a glass with different liquids and layer them so that you can easily see where one ends and the next one begins. The experiment works best if you can find a tall, cone-shaped glass – rather like the glasses used for ice cream in restaurants.

What To Do

First, find some different liquids – surgical spirit, water, wine, oil, and coffee with sugar all work well. Start by pouring in a layer of coffee. When the coffee has settled, carefully add some water, dribbling it down the side of the glass using a small funnel. (If you can't find a funnel, make one out of paper – just remember to make a separate one for each liquid so they don't mix.) Let that settle and then add the other liquids in turn, making sure you let each one settle before you pour in the next one.

How It Works

Each liquid occupies its own layer in the glass because each one has a different density. As long the surface tension of each one is maintained (i.e. as long as you don't just pour in the next layer straight on top of the previous one), they will remain separate. The different colour of each liquid allows you to see the various layers clearly.

92 FINDING YOUR BLIND SPOT

Everyone has a blind spot, but they're all slightly different. This simplest of experiments can help you discover exactly where yours is, and requires almost zero preparation and no fancy equipment.

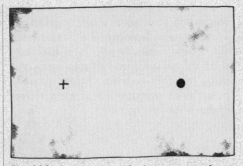

Fig. 119. *As you move the image towards you, the dot on the right hand side will just disappear.*

What You Need
In fact, you're already holding everything that you need in your hands: this book (or, more specifically, the illustration on this page). That's it.

What To Do
You just need to hold the book about 25cm (10in) in front of you, parallel with your eyes. Close your left eye and look at the '+' sign on the left. Don't look at the black dot to the right of it, which you should try to ignore. Then – very slowly – start to move the book towards your nose. Although you're looking at the '+' sign, you're still aware of the black circle, but at some point, as you move the page towards you, it will simply disappear. There won't be any doubt, either – it will literally vanish. If you move the page again or open your other eye or look directly where the spot used to be, you'll see it again.

How It Works
Each of your eyes has a design quirk called a blind spot – it's right at the back of the retina where the optic nerve leaves the eye. Most of the retina in the eye is covered with things called photoreceptors, which grab light and convert it into nerve signals that are transmitted to the brain down miniature 'cables' called axons. At the back of the retina, these axons are actually in front

of the photoreceptors instead of behind them, effectively blocking the light and producing a literal blind spot. As you move the book towards your nose, there comes a point when the light reflecting off the black circle is pointing at the part of the retina that doesn't have any photoreceptors. Hence, you can't see it.

Fig. 120. *It's not really an illusion – so far as your eye is concerned, the black spot literally isn't there.*

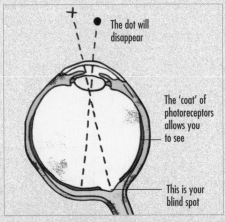

The dot will disappear

The 'coat' of photoreceptors allows you to see

This is your blind spot

93 AUTOMATIC PINBALL

Traditional pinball machines are just too much like hard work – all those springs, and having to pull the handle back and... well, let's just say that all this complexity is unnecessary, as this extraordinary experiment with magnetism demonstrates quite beautifully.

You'll need four strong NIB magnets for this

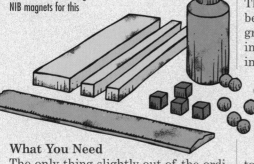

What You Need

The only thing slightly out of the ordinary that's required for this is some very strong magnets – since the whole experiment is powered by magnetism, the stronger they are, the better. The ones to look for are called 'neodymium iron boron' magnets (sometimes called 'rare earth' magnets), which are many times more powerful than conventional ones. After that, all you need are three pieces of wood for the runner, some tape to keep the magnets in place, a pair of scissors, a ruler and a number of round steel balls.

Fig. 121. *Here's a cross section of the three pieces of wood that make the 'track'.*

What To Do

In this experiment we're going to be building a 30cm (12in) track made of wood. For that you'll need three straight lengths: one for the base and two narrower bits which you can glue on the top, leaving a gap in the middle. The base should be 32mm (1¼in) across and the two other bits 13mm (½in) across. This will leave a gap of 6mm (¼in) in between them, which will provide a groove for the balls that will keep them in line (see illustration). Glue the wood in place and then let it dry. Then, gently sand the top inner edge of the track to remove any roughness. The alternative, by the way, is to produce an actual groove in a length of wood, which is easy enough to do if you've got access to a router.

The next step is to tape the magnets onto the track. The magnets are 13mm (½in) square, and since the track is 30cm (12in) long, it makes sense to position the magnets every 5cm (2in). Tear off enough tape to secure the first magnet and then cut it in half lengthways with the scissors. If you think this sounds awkward, then you're right, but it's much easier than taping the magnet down and then trying to trim the excess tape off with metal scissors – those magnets are very strong and will make it very hard work! Once the first magnet is secured, do the same for the second one,

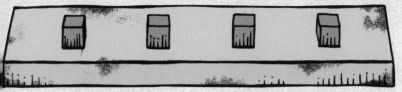

Fig. 122. *Here are the four strong magnets in position – you can use sticky tape or strong glue if you don't need to re-use them for something else.*

but, first, put something in between the two magnets – it won't stop the attraction, but it will make it easier to keep them apart. Repeat until you've positioned all four magnets.

Your automatic pinball is now nearly ready. Keep one ball in your pocket and then place the others in pairs to the right of each magnet and let them sit there. Get a cushion and position it about 3m (10ft) in front of the last ball. Take the final ball from your pocket and place it on the track right at the other end. You don't need to roll it – just let go. If you've set your pinball up correctly, the ball at the far end will shoot off and hit the cushion, and it'll happen so fast that you won't actually be able to see it.

How It Works

It's a combination of kinetic energy and magnetism. The first ball has no speed of its own – you just let it go and it's attracted to the first magnet. When it hits, there's a knock-on effect through the magnet to the ball on the other side. This hits the ball next to it, which sets off at a greater speed than the first ball (i.e. the very first ball which you placed at the end of the track and simply let go of). By the time it hits the second magnet it's travelling much faster than the first ball and transfers that kinetic energy through the second magnet to the ball on the other side, which hits the ball next to it. This sets off in turn, going even faster, and so on. The final ball travels at such a speed that it's able to leap off the track and into the cushion.

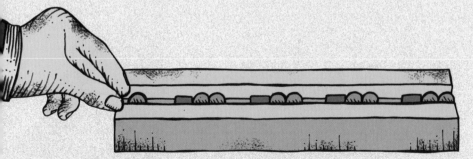

Fig. 123. *You don't need to flick the first ball or anything like that; simply let it go and then let the magnets – and kinetic energy – do their jobs.*

94 PHOTOS IN 3D

This simple experiment will allow you to view your photographs as dramatic three-dimensional images. Just the job for the aspiring landscape photographer.

What You Need

A good solid cardboard box about 50 x 50cm (20 x 20in) and about 15cm (6in) high. It doesn't need to have a lid, but if it does, keep it, because you'll also want some extra bits of cardboard at some stage. Then you'll need some of that blue putty stuff that's good for sticking things on walls, a ruler, pencil, sharp craft knife, some glue and a four mirrors – two of these need to be about 15cm (6in) square and the other two about half that size. You may be able to find mirrors like this already cut to size, but if you can't a local glazier will do it for you.

What To Do

First you need to cut a hole in the middle of one side of your box. The hole needs to be 15 x 5cm (6 x 2in), and you can use the ruler to position it correctly. Use the craft knife to make the hole, and run tape around the edges to smooth it off nicely. You now need to cut out four squares of cardboard measuring 15 x 15cm (6 x 6in). Cut one of these in half across the diagonal, so you end up with two triangles. Take one of the other squares and score it carefully down the middle with the craft knife to make it easier to fold. Then take the ruler, lay it along the line you just scored, and carefully fold the cardboard along the line. You can now glue the folded piece of cardboard to the edge of

one of the triangles so it looks like the prow of a ship (see middle illustration).

Put a dot of the blue putty stuff on each corner of the smaller pair of mirrors and then attach them to the corner 'wall' you just created. (When the mirrors are in exactly the right position, you can glue them down if you want.) Next, measure about 13cm (5in) up one side of the box on the inside, from the end where you cut the viewfinder hole. Make a mark there with a pencil, then do the same on the other side. Take your two remaining squares of cardboard, stick the last two mirrors to them with the putty, and place them in the bottom corners of the box – nearest the viewfinder hole – at right angles to the corners, with the mirrors facing into the box. Position the double 'V' mirror you made previously as shown in the picture. Later, these will need to be positioned so that light from photos at the back end of the box can bounce off the corner mirrors, onto the 'V' mirrors and into your eyes.

Fig. 124. *Start by taking a photo of something you love – like your shed – then step to the side and take the same photo again.*

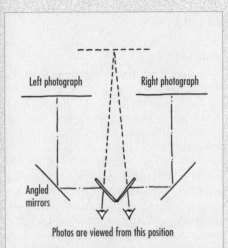

Fig. 125. *Here's the plan view showing the position of the photographs at the back and the mirrors at the front of the box.*

side without moving your feet. This will alter the angle just enough to let you get the second shot. Develop the pictures, or print them out yourself, and then stick them to the back of the box with the first photo on the right and the second photo on the left. Sit in front of the viewfinder hole and adjust the mirrors as necessary in order to see the photos reflected in the mirrors. Amazingly, they will appear to form a three-dimensional image.

How It Works

Look at an object and then close one eye; then open it and close the other. Keep looking at the object and opening and closing each eye in turn. You'll start to become aware that each eye sees a slightly different view of the same object. The 3D viewer simply puts these two images together and encourages your eyes to view them as a single three-dimensional object.

Next, you'll need a couple of pictures to actually look at. Here the trick is to start with something large and obvious – how about your garden shed? Start by taking a few photos with the shed clearly in the foreground and with something easy to spot (like a door or a window) in the middle of the shot. Then, take two ordinary steps to the left and shoot the scene again, making sure that the same thing is in the centre of each photo. This works well when the object you're photographing is a little way away from you. However, if you want to photograph something in the same room – for example a brightly coloured vase of flowers – then you'll need to decrease the distance between the two photographs. For something close up, instead of taking actual steps, try rocking to one

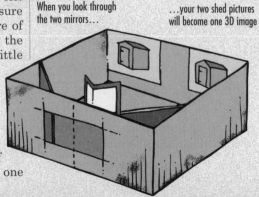

When you look through the two mirrors...

...your two shed pictures will become one 3D image

115

95 MAKE SQUARE BUBBLES

Bubbles are round, right? From the time you used to blow them in your buggy, using nothing but natural materials, they've always been round... unless they're square.

What You Need

Some drinking straws, some modelling clay, something to blow bubbles with (perhaps a plastic biro with the middle taken out, or a left-over shop-bought bubble blower), a bowl large enough to dip the straw cube into, and a supply of bubbles.

What To Do

Start by making a cube with the straws and modelling clay. Simply roll the clay between your palms to make eight balls of equal size and then press the straw ends into them to form the cube – as long as the straws are the same length, it's easy. Use a mixture of water and washing-up liquid to make the bubbles. Start with the water and then keep adding the detergent, testing with the blower until you get consistently good bubbles. Now, dip the whole of the straw cube into the liquid, leave it there for a second, remove it and give it a gentle shake. The bubble sides will snap into the middle to form an hourglass shape. Finally, blow a small bubble and then tap the blower so the bubble falls into the middle of the hourglass. Give the straw box another gentle shake and you'll see a square bubble in the middle.

How It Works

A bubble is normally round because it's trying to occupy the least amount of surface area relative to its volume. When you lower the straw box into the solution, the same thing happens – the bubble solution produces six films that hug the sides of the cube. However, because the bubble molecules are drawn together, when you shake the cube, they're attracted and form the hourglass shape. When you add the bubble, its surface tension is strong enough to hold its shape, but not so strong that it breaks the other bubble films. In fact, if you look carefully, your square bubble is desperately trying to make a sphere again.

Fig. 126. Use straws and modelling clay to make your frame and then dip it in the bowl of bubble liquid.

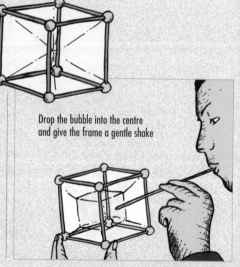

Drop the bubble into the centre and give the frame a gentle shake

96 HOW ARE THUNDERSTORMS FORMED?

The power of a thunderstorm is awesome. Fortunately, this little experiment shows you exactly how thunderstorms happen without generating a single spark. All you need is a shoebox-sized, see-through waterproof container, some blue and red food colouring and a way of making ice cubes.

What To Do

Start by squirting a few drops of blue food colouring into some water and then giving it a good mix. Pour it into a tray of ice cubes and stick them in the freezer. When they're done, fill the container about two-thirds full of warm water. After a moment or two, pop a couple of ice cubes into the water at one end and then add a couple of drops of red food colouring at the other end of the container. Observe how the water in the container divides into two layers.

How It Works

Thanks to convection, warm stuff rises and cold stuff sinks, and that's what's happening here. Imagine that the blue water is a cold front and that the red is a mass of warmer air. The cold air forces the warm air to rise until it cools off and condenses, producing water vapour.

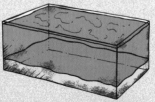

A thunderstorm in miniature – only not so wet and noisy

97 THE SPINNING EGGS

Where would be without eggs? Lovely to eat and – more to the point – great for experiments like this one, which illustrates Newton's First Law.

What To Do

All you need is a couple of eggs. Leave one uncooked and boil the other one so that it's hardboiled. Then, all you have to do is spin the eggs in turn on a smooth surface where the eggs aren't going to spin off and break (obviously, this is especially important for the uncooked egg). Give the cooked egg a good spin and then stop it spinning and take your hand away immediately. Watch what happens. Now repeat the experiment with the uncooked egg. Although both eggs look the same, you'll see that different things are happening when you spin them.

How It Works

Newton's First Law states that when an object moves, it continues at a steady speed until it's pushed or pulled by an outside force. Because the yolk and the white of the uncooked egg aren't attached to the shell, they continue to move even after you've stopped the egg.

98 OPTICAL ILLUSIONS

You've probably heard someone moaning that they can't trust their own eyes any more. Well, if you show them these two pages, they'll probably be right. Here's a great selection of unusual visual twisters that will make you think you're seeing things when you aren't.

What You Need
Only the illustrations on these two pages, a bit of patience and some decent – preferably natural – light. Apart from that it's just you, your eyes and the puzzles.

How It Works
In each case, you can either read the explanation that accompanies each visual puzzle first, or check the puzzle itself out and try to work out for yourself what's going on. There's no right way to do it; just suit yourself. After you've looked over the puzzles, you might want to try them out on a friend, in which case you can decide for them.

COUNT THE CUBES
There are nine cubes in this picture – correct? Well, that's what most people see because we usually take our orientation from the top cube and the easiest way to see that is with the upper diamond shape forming the top. But if you look at the image for a little longer, you'll see that there are various ways to look at it: as four 3D cubes, or as a series of inverted cubes. Try rotating it 90 degrees either way, for example, and see what happens.

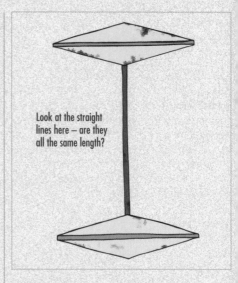

Look at the straight lines here – are they all the same length?

SAME LENGTH
The three lines in this picture are all different lengths, right? Wrong. The vertical one in the middle is longer than the other two. Wrong again. If you measure them, you'll discover that they're all the same length. Because the two elongated diamonds at the end of the vertical line slope away from it, they kid your eye into thinking that the vertical line is longer.

Look carefully at this picture and you'll see nine cubes...

...or are there actually only four?

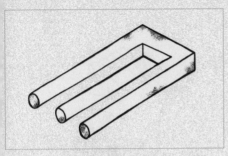

SHAKY SIGHT

If you stare at this shape for any length of time it starts to shiver and shake, particularly if you start at the centre and then look about halfway out to the edge. The thing moving isn't the image but your own eye, which flickers back and forth constantly, and so quickly that most of the time you aren't even aware of it.

IMPOSSIBLE CRICKET STUMPS

At first sight, it's a fairly ordinary pair of cricket stumps, but on close examination you realise that this is an Escheresque nightmare of impossible angles and planes. It's exactly that. Our brain takes the flat image and tries to turn it into a three dimensional one that would make sense in the physical world. Trace the lines with your finger, however, and you can see that this is impossible.

CROOKED LINES

At first glance, the diagonal lines in this picture look disjointed, as if there's no way on earth that they'd line up. However, in actual fact, they're not crooked at all. If you take a ruler and lay it along the smaller lines crossing the thicker vertical lines, you'll find that they line up perfectly. The thicker lines confuse your brain into trying to link up the wrong diagonal lines, and that's what causes the illusion.

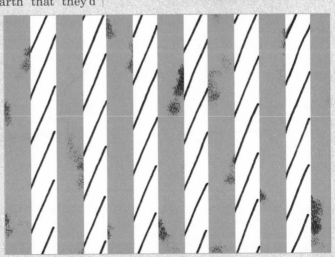

99 THE FLYING TEA BAG

There are many ways to make a tea bag fly. You can throw it at someone if you like – that's a kind of flying – or you can follow this experiment, and watch as a deconstructed tea bag shoots straight up in the air.

What You Need

A tea bag – one of the ones on a string works best because of the way the bags are constructed (round bags are hopeless because they're the wrong shape). You'll also need a pair of scissors and a box of matches. Don't attempt this indoors, because the flight of a burning tea bag is unpredictable – try it outdoors on a windless day.

What To Do

Cut off the top of the tea bag, just underneath the staple, and chuck away the tea (unless you still own a teapot, in which case you can brew up while you're doing this experiment). Carefully unfold what's left of the tea bag and you'll discover that you've got a nice little tube made of this flimsy tea bag material (which is why you need a windless day). Pop your finger inside the tube and open it out. Then stand it on a flat surface and light the top with a match. It will burn just as you'd expect until it gets about halfway down, when the whole lot will suddenly lift off and sail away into the air.

Cut the top off and empty the contents

You'll believe a tea bag can fly!

How It Works

This is a good example of expansion doing its stuff. Heated by the flames, the air inside the tea bag expands, making it lighter and allowing it to lift off. If you don't have access to any tea bags, you can achieve a similar result by taking a piece of paper, rolling it into a tube and then twirling the end into a point. Light the pointed end and watch what happens.

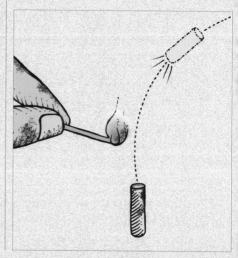

Fig. 127. *When you light the top of the tea bag it'll burn quickly and then suddenly lift off into the air.*

100 THE ESCAPING PLANT

Here's a simple experiment to show that a plant will always seek the light in order to grow and remain healthy. It's straightforward to set up and you can leave it for a few days to 'escape' on its own.

What You Need

A small flowerpot, a bit of moist soil, a potato that's starting to sprout (then no one will miss it from the kitchen), an oblong box of about 30.5–46cm (12–18in) long by 15cm (6in) wide by a little more than the height of the plant pot high. If the potato and pot are small enough, an old shoe box will do the trick. You'll also need a sheet of stiff cardboard to make the partitions.

What To Do

Pop the sprouting potato into the pot and cover it with soil. Lift the lid off the box and measure the inner dimensions, then cut out three similarly shaped pieces from the cardboard – these will be used to create the partitions in the box.

Check that the cardboard partitions fit, and then use a pair of sharp scissors to cut a small (slightly larger than potato-sprout size) hole in each one. To give the potato a challenge, cut one hole near one end of the first partition, the second one at the opposite end of the second partition, and the third at the opposite end of the third partition. Make a final hole in one end of the box.

Put the plant pot in the corner furthest away from the end of the box that has a hole in it, then slot the partitions into place and put the lid

back on, making sure no light can get in. Put the box in the sunlight so that light shines in through the hole at the end. Leave it, and after a few days you'll see the shoot poking out through the hole in the end of the box.

How It Works

It's to do with 'tropism', which describes the way a plant responds – by moving or turning – to an external stimulus. For example, the roots of a plant grow towards water, and this is called 'hydrotropism'. In this experiment, however, the shoots grow towards the light so that they can photosynthesise the sunlight into the fuel they need to grow – and this is called heliotropism. Some flowers, for example, actually follow the path of the sun from east to west during the course of the day. Here, the light-sensitive cells in the sprouting potato are able to sense where the light is strongest and then weave their way through the various holes and into the light.

Fig. 128. *Given time, the sprouting potato will weave its way through the holes in the box, seeking sunlight in order to thrive.*

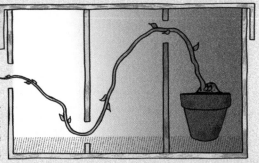

101 THE SOLAR SYSTEM IN YOUR SHED

And finally, we come to the last experiment in the book. It seems fitting for shed scientists to consider the wonder that is the solar system, perhaps better to understand their place within it. Certainly, the way we view our neighbours in space has changed dramatically, as these two simple experiments show.

What You Need
A piece of cardboard about 30cm (12in) square, some stiff paper (in a selection of different colours) from a craft shop. You'll also need a drawing compass, a ruler, some thread, a pair of scissors, some string, a roll of tape, a drawing pin and a nail. Finally, for the second part of the experiment you'll need a roll of toilet paper and a marker pen.

Fig. 129. *The solar system as envisaged by Copernicus only had six planets, but it did have the sun at its centre.*

What To Do
We're going to start by making the solar system as envisaged by Copernicus (though of course he didn't know about all of the planets back then). Start by taking the square of cardboard and using the ruler to find the centre. Tie one end of the thread round the drawing pin and stick it into the centre. Take the other end of the thread and tie it round the pencil. Rotate the pencil round the cardboard to draw out your circle, then cut it out with the scissors.

Take the compass and draw out circular orbits for the four planets closest to the sun. Leave a gap after number four (Mars), and then get your drawing-pin-thread-and-pencil combo and draw out the orbits of the last five planets. Take the nail and enlarge the hole in the centre for the Sun, then make one hole in each of the nine circles – for the purposes of this model, it doesn't matter where you make the holes, as long as they don't form a straight line. Use the compass to draw out a series of circles on the different-coloured cards; the Sun should be the biggest, followed by Jupiter, then Saturn, Uranus, Neptune, Earth, Venus, Mars, Mercury and Pluto.

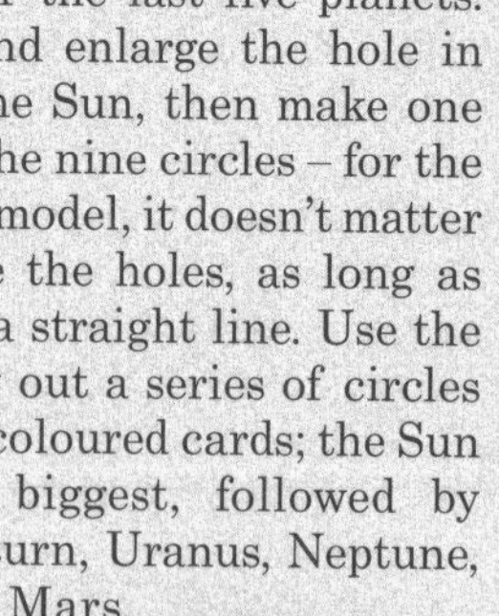

Poke holes in the cardboard for each planet

Fortunately, the orbits don't have to be precise

Fig. 130. *We know that there are more planets in the solar system than Copernicus thought, so be sure to make enough holes.*

Cut off nine equal lengths of thread, poke holes in the top of each heavenly body, and tie on the thread. Poke the other end of the thread up through the relevant hole (see illustration for the correct order of the planets) and tape it in place. Finally, secure the model to the ceiling using nails or drawing pins.

*Fig. 131. **Remember, this is just a representation of the solar system with the sun in the middle.***

So that's the broad Copernican view of the solar system, but today we also understand the enormous distance between the Sun and Pluto, the outermost planet. You can demonstrate that with the roll of toilet paper. Sit in your shed and pick an object to play the Sun (you can even be the Sun yourself). Take hold of one end of the toilet paper and get a helper to start unrolling it. Get them to stop when they've unrolled one sheet – they can use the marker pen to indicate Mercury's position. At 1.8 sheets they can mark Venus; at 2.5 the Earth; at 3.8 Mars; at 13.2 it's Jupiter; at 24.2 sheets it's time for Saturn. Get them to move more carefully now because Uranus is at 48.6 sheets; Neptune at 76.3; and Pluto is a whopping 100 sheets away.

How It Works

The toilet-paper solar system is based on a simple measurement – that one piece of toilet paper is the equivalent to roughly 58 million kilometres (about 35.5 million miles). The toilet-paper technique works well because you can prepare it beforehand, it doesn't require any tape measures, it's portable – and when you've finished with it, the roll can still be used for its original purpose.

Nicolaus Copernicus

Copernicus was the modern European astronomer who first posited the theory that all the planets in the solar system orbit the Sun, and that the Earth was not the centre of the Universe. Given that he had absolutely no experimental evidence with which to back up his claims, this was pretty show-stopping stuff. In fact, Copernicus wasn't certain how it worked and was so nervous about his theory that he didn't publish it until he knew he was dying. It would be more than 100 years before Galileo – thanks to the recently invented telescope – became convinced he could demonstrate that the Earth revolved around the Sun.

*Fig. 132. **Yes, it's possibly the first time a roll of loo paper has been taken out to a wooden building in the garden and used for scientific purposes – in this instance, accurately measuring distances in our solar system.***

GLOSSARY OF TERMS

Absorption What happens when one substance permeates another.

Acid Water-soluble compound that reacts with a base to form salt.

Alkali A kind of chemical base.

Atom The basic component of all matter.

Axon The 'wire' that transmits information out of a nerve cell.

Base Broadly, the opposite of an acid.

Capillary action The way surface tension helps water rise despite the force of gravity.

Catalyst Something that encourages a chemical reaction without being part of it.

Cell The most basic 'unit' that can exist by itself as a living thing.

Charge A characteristic of some particles, either positive or negative.

Chromatography A way of separating complex chemicals into their constituent parts.

Circuit A path followed by electricity.

Concentrate A solution that contains a large amount of solute.

Condense What happens when a gas is cooled and turns into a liquid.

Conductor Something able to transmit electricity or heat.

Convection The vertical transport of heat and moisture.

Copernicus Astronomer who posited that the earth moved round a stationary sun.

Crystal radio Primitive radio that uses the radio signal itself for power.

Current Electricity that flows through a conductor or around a circuit.

Density The ratio of the mass of a substance to the volume it occupies.

Diode Device that lets current flow in a single direction only.

Distilled Usually refers to water that's been treated to remove minerals and other impurities.

DNA Molecule that encodes the genetic information in the nucleus of cells.

Electrode Conductive component through which electricity flows in or out.

Electrolysis Chemical energy produced when a current passes through a liquid.

Electron A negatively charged particle.

Enzyme A protein that speeds up chemical reactions.

Frequency The number of sound vibrations in a second.

Friction Resistance experienced when one body is in contact with another.

Galileo Pioneer of the scientific method; used a telescope to prove the earth moved round the sun.

Gravity Force of attraction shared among everything that has mass.

Hydrotropism Tendency of a plant's roots to grow in the direction of water.

Ionization Process whereby atoms acquire a positive or negative charge.

Kinetic energy Produced when objects bang into each other.

LED Light-emitting diode.

Molecule Smallest unit of a compound that exhibits all the properties of that compound; larger than an atom.

Multimeter Device that can be set to measure volts, ohms, and amperes or milliamperes.

Neodymium iron boron (NIB) Compound used to produce a magnet up to 10 times stronger than traditional ferrite magnets.

Neutral Neither acid nor alkaline.

Osmosis The movement of a solution through a membrane so as to equalise the concentrations on both sides.

Oxidation Chemical reaction which occurs when something reacts with oxygen.

Permeate Spreading or diffusing through something.

pH indicator Measures acidity or alkalinity of something dissolved in a solution; expressed as a number between 1 and 14.

Photoreceptor A cell that's sensitive to light.

Photosynthesis The way plants produce carbohydrates and oxygen, courtesy of chlorophyll and sunlight.

Precipitate What you get when a substance separates from a solution because of a chemical reaction.

Pressure Force produced by pressing on a surface.

Reaction Chemical transformation when two or more substances interact.

Refraction The way light bends when it moves from one medium to another (for example, between air and glass).

Resistance The way something opposes the flow of an electrical current.

Resonance The way things vibrate with each other.

Solute Anything that is dissolved in a solution.

Solution Mixture of dissolved substances.

Static The kind of electricity you get when electrons are encouraged to jump from one atom to another.

Supersaturation A solution that's been heated so as to dissolve more solute than usual.

Surface tension The way molecules are attracted to each other on the surface of a liquid to produce a 'skin'.

Transpiration Describes the way plants pull in water from their roots and release it as water vapour through their leaves.

Vacuum The removal of air from a space so it drops below atmospheric pressure.

Variable transformer Allows fine control over voltage supplied to electrical equipment.

INDEX

Project 84
Measuring Wind
Speed, page 101

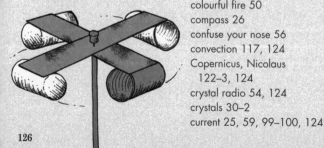

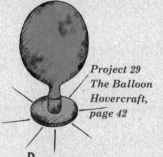

Project 29
The Balloon
Hovercraft,
page 42

*Project 85
Fireproofing,
page 102*

*Project 49
Turn a Rainbow
White, page 64*

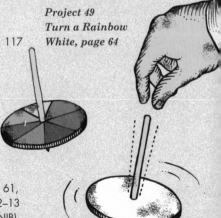

127

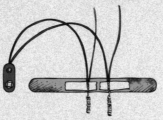

Project 82 Make a Water-Powered Fuel Cell, page 98

Project 79 The Can Crusher, page 60